The Principles and Practice of
Canal and River Engineering

DAVID STEVENSON

CAMBRIDGE
UNIVERSITY PRESS

CAMBRIDGE UNIVERSITY PRESS

Cambridge, New York, Melbourne, Madrid, Cape Town,
Singapore, São Paolo, Delhi, Mexico City

Published in the United States of America by Cambridge University Press, New York

www.cambridge.org
Information on this title: www.cambridge.org/9781108057721

© in this compilation Cambridge University Press 2013

This edition first published 1872
This digitally printed version 2013

ISBN 978-1-108-05772-1 Paperback

This book reproduces the text of the original edition. The content and language reflect
the beliefs, practices and terminology of their time, and have not been updated.

Cambridge University Press wishes to make clear that the book, unless originally published
by Cambridge, is not being republished by, in association or collaboration with, or
with the endorsement or approval of, the original publisher or its successors in title.

CAMBRIDGE LIBRARY COLLECTION

Books of enduring scholarly value

Technology

The focus of this series is engineering, broadly construed. It covers technological innovation from a range of periods and cultures, but centres on the technological achievements of the industrial era in the West, particularly in the nineteenth century, as understood by their contemporaries. Infrastructure is one major focus, covering the building of railways and canals, bridges and tunnels, land drainage, the laying of submarine cables, and the construction of docks and lighthouses. Other key topics include developments in industrial and manufacturing fields such as mining technology, the production of iron and steel, the use of steam power, and chemical processes such as photography and textile dyes.

The Principles and Practice of Canal and River Engineering

One of the leading figures in the age of great engineers, David Stevenson (1815–86) was the son of a lighthouse builder, and while studying at Edinburgh University he was already gaining experience at his father's side. It is for his lighthouses and works of inland navigation that he is best remembered: he designed Britain's most northerly lighthouse and worked on improving navigation on rivers such as the Dee, the Forth and the Clyde. His article on inland navigation for the *Encyclopaedia Britannica* was published separately in 1858, and was revised and updated for this second edition in 1872. Although Stevenson acknowledges that the age of the canal has been superseded by the age of the railway, he maintains that there is much to learn from the older technology. Illustrated with cross-sections and plans, this work will be of interest to readers seeking to explore the history of Britain's industrial infrastructure.

Cambridge University Press has long been a pioneer in the reissuing of out-of-print titles from its own backlist, producing digital reprints of books that are still sought after by scholars and students but could not be reprinted economically using traditional technology. The Cambridge Library Collection extends this activity to a wider range of books which are still of importance to researchers and professionals, either for the source material they contain, or as landmarks in the history of their academic discipline.

Drawing from the world-renowned collections in the Cambridge University Library and other partner libraries, and guided by the advice of experts in each subject area, Cambridge University Press is using state-of-the-art scanning machines in its own Printing House to capture the content of each book selected for inclusion. The files are processed to give a consistently clear, crisp image, and the books finished to the high quality standard for which the Press is recognised around the world. The latest print-on-demand technology ensures that the books will remain available indefinitely, and that orders for single or multiple copies can quickly be supplied.

The Cambridge Library Collection brings back to life books of enduring scholarly value (including out-of-copyright works originally issued by other publishers) across a wide range of disciplines in the humanities and social sciences and in science and technology.

INLAND NAVIGATION.

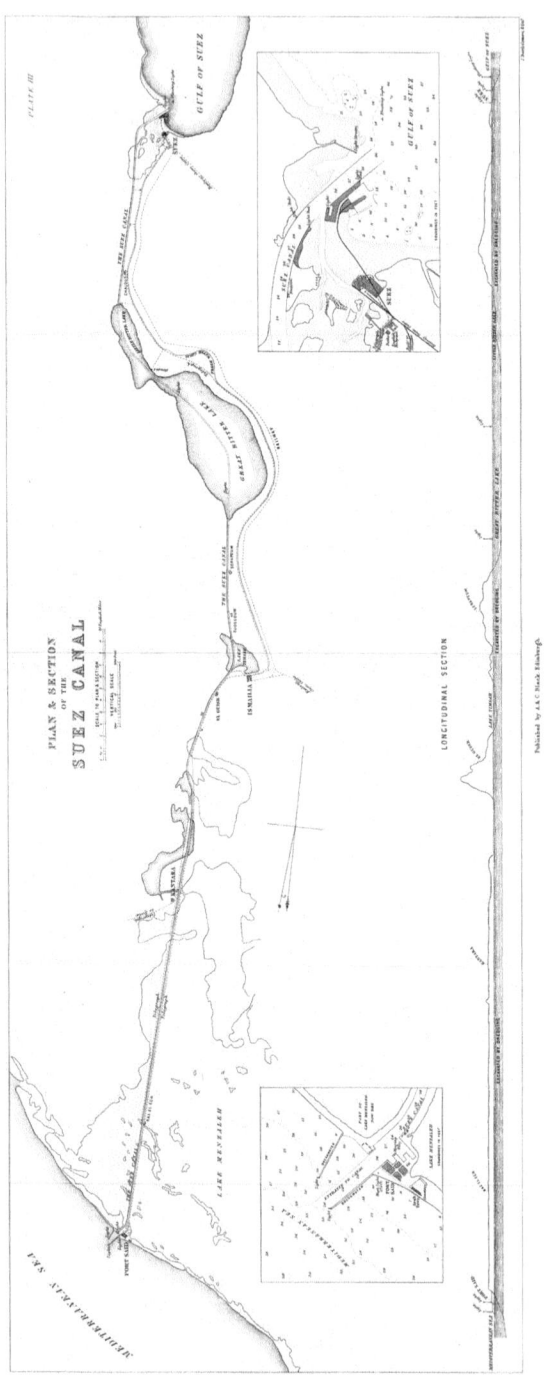

The material originally positioned here is too large for reproduction in this reissue. A PDF can be downloaded from the web address given on page iv of this book, by clicking on 'Resources Available'.

THE

PRINCIPLES AND PRACTICE

OF

CANAL AND RIVER

ENGINEERING

BY

DAVID STEVENSON, F.R.S.E.

MEMBER OF THE INSTITUTION OF CIVIL ENGINEERS;

AUTHOR OF "A SKETCH OF THE CIVIL ENGINEERING OF NORTH AMERICA,"
"TREATISE ON THE APPLICATION OF MARINE SURVEYING AND HYDROMETRY TO THE
PRACTICE OF CIVIL ENGINEERING," ETC.

SECOND EDITION.

EDINBURGH
ADAM AND CHARLES BLACK
1872.

[*The right of Translation reserved.*]

PRINTED BY T. AND A. CONSTABLE, PRINTERS TO HER MAJESTY,
AT THE EDINBURGH UNIVERSITY PRESS.

PREFACE TO SECOND EDITION.

SOME of the information contained in the following Chapters was published as the article "Inland Navigation," in the eighth edition of the *Encyclopædia Britannica*, and was afterwards, in 1858, issued by the publishers as a separate work, which has for some time been out of print.

Messrs. Black have asked me to prepare another edition; and in complying with their request it has been thought right, not only to notice everything that could be regarded as new in Canal and River Engineering, but also to alter and extend the original matter, instead of retaining the condensed style of the first issue, which was more suitable for the columns of an encyclopædia than the pages of an independent treatise.

The reader will find in the first two chapters a sketch of the early history of Barge Canals, and a statement of the chief features of their construction, followed by a notice of those larger works used by sea-borne vessels, which are divided into three distinct classes, represented by the Caledonian, the Amsterdam, and the Suez Canals, of which I have given a description.

The second part of the subject, relating to Rivers and
Estuaries, embraces a much wider field of inquiry, and
includes some of the most difficult problems with which
the engineer has to deal. Before entering on it, I thought
it desirable to explain certain operations in Hydrology, in
order to enable the student of Engineering to follow with
more advantage the succeeding chapters. These explanations refer chiefly to the method of making tidal observations, and ascertaining the discharge of streams.

All rivers on their passage to the sea through tidal
channels and estuaries, assume certain recognisable and
widely different physical features, the characteristics and
boundaries of which I have defined. These boundaries
divide rivers into what, for our present purpose, may
perhaps conveniently be termed *engineering compartments;* and I have explained the varied works which are
applicable to each of them, and shown their effects by
describing different navigations where they have been
successfully adopted.

The concluding chapters are chiefly devoted to the
origin of "Bars," the effect of "Backwater," and the
Reclamation and Conservation of Land.

It seems to me that such branches of Engineering
may be most usefully discussed by explaining general
principles,—describing works designed to produce certain
effects,—showing their application in practice, and record-
ing results. This is the plan I have endeavoured to
follow, and accordingly every statement of principle has

been illustrated by at least one example in practice. Where my own experience failed to suggest examples, I had no difficulty in applying to my professional brethren, and I have great satisfaction in acknowledging the friendly interest they took in assisting me to carry out the object I had in view.

It will thus be found that the following pages contain a *résumé* of a pretty wide field of research, which I trust may prove in some respects useful, if not to the engineer in his practice, at least to the pupil in the study of his profession.

EDINBURGH, *April* 1872.

CONTENTS.

CHAPTER I.

BARGE CANALS.

CHAPTER II.

SHIP CANALS.

CHAPTER III.

THE COMPARTMENTS OF RIVERS DEFINED.

CHAPTER IV.

HYDROMETRIC OBSERVATIONS.

CHAPTER V.

DISCHARGE OF RIVERS—UNDER-CURRENTS—SPECIFIC GRAVITIES OF WATER, ETC.

CHAPTER VI.

THE "RIVER PROPER" COMPARTMENT.

CHAPTER VII.

TIDAL PROPAGATION AND TIDAL CURRENTS OF RIVERS.

CHAPTER VIII.

TIDAL COMPARTMENT—WORKS FOR ITS IMPROVEMENT.

CHAPTER IX.

APPLICATION OF THESE WORKS IN PRACTICE.

CHAPTER X.

CHAPTER XI.

WORKS FOR ACCOMMODATION OF VESSELS.

CHAPTER XII.

"SEA PROPER" COMPARTMENT OF RIVERS.

CHAPTER XIII.

RECLAMATION AND PROTECTION OF LAND.

CHAPTER XIV.

LIST OF PLATES.

INLAND NAVIGATION.

CHAPTER I.

BARGE CANALS.

Early history of Barge canals—Invention of locks—Languedoc Canal—Foss Dyke
and Caer Dyke Canals—Bridgewater and other Canals—Difficulties in constructing early canals—General principles of canal construction—Supply of water
—Sectional area—Reaches and locks—Inclined planes and perpendicular lifts
—Monkland Canal incline—Waste weirs—Stop-gates—Off-lets—Drainage of
towpaths—Puddle—Mode of conducting traffic—Wasting of the banks—
Steam-towing on Gloucester and other canals—Steam-towing on rivers.

THAT railways have entirely superseded, and will in Early history of
Barge canals.
future prevent, the extension of canal or water carriage
as a means of ordinary transport, must at once be con-
ceded. The days of the bargemen baiting their horses at
every stage, and travelling from town to town and from
village to village at three miles an hour, are ended, and
the change is not to be regretted. The great objections
to relying on canals for keeping up regular internal communication are their liability to stoppage from a deficient
supply of water during dry seasons, the interruption to
which they are exposed from ice during winter, and,
especially, in these days of express railway trains and
electric telegraphs, the very limited speed at which the
canal-boats can be propelled. Sir John Rennie, some
twenty years ago, in speaking of the successful attempts

made to introduce swift boats, and the great improvement
that was thereby effected in canal transport, says,—" All
this, however, came too late; for although it would have
been readily acknowledged at an earlier period, and might
perhaps, for a while, have retarded the railway system,
yet when once the latter was established, its superiority
became manifest, and its progress irresistible."[1]   To
modern travellers the old canal track-boat, as compared
to the railway train, seems to have got so completely out
of date, that it may at first sight be considered as uncalled
for to describe, even briefly, a class of works which may
almost be regarded as obsolete.   It appears to me, however, that the simple consideration of the great antiquity
of navigable canals, their wide-spread introduction throughout the world, the important place which they have so
long occupied in the commercial history of every country,
and, above all, the noble specimens which they afford of
hydraulic engineering of a confessedly difficult nature,
executed at a period when the mechanical appliances of
modern times were unknown, give to such works an important place in the history of Inland Navigation.

From the writings of Herodotus, Aristotle, Pliny, and
other ancient historians, we learn that canals existed in
Egypt before the Christian era, and there is reason to
believe that at the same early period artificial inland
navigation had also been introduced into China.   Almost
nothing, however, save their existence, has been recorded
with reference to these very early works; but soon after
the commencement of the Christian era canals were intro-

---

[1] *Minutes of Proceedings of Institution of Civil Engineers*, vol. v. p. 78.

duced, and gradually extended throughout Europe, in Italy, Spain, Russia, Sweden, Holland, and France.[1]

In speaking, however, of the earliest of these works, it is not to be supposed that they resembled the modern canals constructed in our own and other countries. It was not until the invention of canal-locks, by which boats could be transferred from one level to another, that inland navi- Invention of gation was rendered generally applicable and useful; and it locks. has been truly remarked, "that to us, living in an age of steam-engines and daguerreotypes, it might appear strange that an invention so simple in itself as the canal-lock, and founded on properties of fluids little recondite, should have escaped the acuteness of Egypt, Greece, and Rome."[2] Not only, however, had the invention escaped the notice of the ancients, but what is more striking, the several gradations made towards the attainment of that simple but valuable improvement appear to have been so gradual, that, like many discoveries of importance, great doubts exist as to the *person* and even the *nation* by whom canal-locks were first introduced. One class of writers attributes the discovery to the Dutch; and Messrs. Telford and Nimmo, who wrote the article on Inland Navigation in Brewster's *Edinburgh Encyclopædia*, adopt the conclusion that locks were used in Holland nearly a century before their application in Italy; while, on the other hand, the invention has been strongly and not unreasonably claimed for engineers of the Italian school,

[1] Fulton on Canal Navigation, London, 1796 ; Vallancey's *Treatise on Inland Navigation*, Dublin, 1763; Tatham's *Political Economy of Inland Navigation*, London, 1799; "Inland Navigation," Brewster's *Edinburgh Encyclopædia*.
[2] *Quarterly Review*, No. cxlvi., "Navigable Canals," by Paul Frisi.

and, in particular, for Leonardo da Vinci, the celebrated engineer and painter.[1] Without, however, entering into a discussion of this question, which it is now perhaps impossible to solve, we may safely state, that during the fourteenth century the introduction of locks, whether of Dutch or Italian origin, gave a new character to inland navigation, and laid the basis of its rapid and successful extension. And here it may be proper to remark, that the early canals of China and Egypt, although destitute of locks, do not appear to have been on that account formed on a uniform level unadapted to varying heights. I do not know that the use of locks has even yet been introduced into China, intersected as it is by many canals of great antiquity and extent;[2] and in order to pass boats from one level to another, the Chinese have, from a very early period, employed stop-gates and inclined planes of rude construction. Nevertheless, the invention of locks was, as already noticed, a most important step in the history of canals; and that mode of surmounting elevations may be said to be almost universally adopted throughout Europe and America. Inclined planes and perpendicular lifts have, it is true, been employed in these countries, as will be noticed hereafter; but the instances of their application are undoubtedly rare.

Languedoc Canal.

But without tracing the gradual introduction of canals from country to country, I remark at once, that we find the French, at the end of the seventeenth century, in

---

[1] Frisi on Navigable Canals, p. 154.

[2] The Imperial Canal in China is said to be upwards of 1000 miles in length.

the reign of Louis XIV., forming the Languedoc Canal, between the Bay of Biscay and the Mediterranean, a gigantic work, which was finished in 1681. It is 148 miles in length, and the summit-level is 600 feet above the sea; while the works on its line embrace upwards of one hundred locks and about fifty aqueducts,—the whole forming an undertaking which is a lasting monument of the skill and enterprise of its projectors; and with this work as a model, it seems strange that Britain should not, till nearly a century after its completion, have engaged in vigorously following so laudable an example. But this indeed is all the more extraordinary, seeing that the Romans in early times had executed works in England, which, whatever might have been their original use, whether for the purposes of navigation or drainage, were ultimately, and that even at an early period, converted into navigable canals. Of these works I particularly specify the Caer Dyke and Foss Dyke cuts in Lincolnshire, which are by general consent admitted to have been of Roman origin. The former extends from Peterborough to the river Witham, near the city of Lincoln, a distance of about 40 miles; and the latter extends from Lincoln to the river Trent, near Torksey, a distance of 11 miles.

Of the Caer Dyke the name only now remains, but the Foss Dyke, though of Roman origin, still exists, and, as it is the *oldest British canal*, the reader may be interested to learn the following facts as to its history, which I gathered some years ago, when designing works for its improvement. Camden in his *Britannia* states that

Foss Dyke Canal.

the Foss Dyke was a cut originally made by the
Romans, probably for water-supply or drainage, and
that it was deepened and rendered in some measure
navigable in the year 1121, by Henry I. In 1762 it was
reported on by Smeaton and Grundy, who found the depth
at that time to be about 2 feet 8 inches.[1] They, however, discouraged the idea of deepening it by excavation.
They say they found "the bottom to be either a rotten
peat earth, or else a running sand," and that though the
deepening of the navigation is in "nature possible," yet
it "cannot be effected without removing one of the banks,
in order to widen the same," which would not only turn
out expensive, but would "occasion much loss of time
and profit to the proprietor, while the work is executing."
Nothing followed on this report, but in 1782 Smeaton
was again called in, and deepened the navigation to 3 feet
6 inches, not, however, by widening the canal or dredging,
but by raising the water-level 10 inches.[2] From that
period nothing more was done to enlarge the water-way,
or adapt it to increased traffic. Meantime, however,
the adjoining Witham navigation, having been improved, the defects of old Foss became more apparent,
and in 1838 Mr. Vignoles was consulted, and made an
elaborate report on alternative schemes for increasing the
depth to 4 and 6 feet; nothing, however, was done till
1840, when Messrs. Stevenson of Edinburgh were employed to design works for assimilating the Foss Dyke, as
far as practicable, both as regards width and depth, to the
navigable channel of the Witham. On examination I

---

[1] Smeaton's *Reports*, vol. i. p. 55: London, 1812.   [2] *Ibid.* p. 74.

found the depth of the Foss to be 3 feet 10 inches, and its breadth in many places was insufficient to admit of two boats passing each other, and for their convenience occasional passing places had been provided. It was resolved to increase the dimensions of the canal, and to repair the whole work. Accordingly it was widened to the *minimum* breadth of 45 feet, and deepened to the extent of 6 feet throughout; alterations which were accomplished, without stopping the traffic, by using steam and hand dredges, specially designed for the purpose. The entrance-lock, communicating with the river Trent at Torksey, was renewed, and a pumping-engine was erected for supplying water from the Trent during dry seasons, and thus that ancient canal, which is quoted by Telford and Nimmo as "the oldest artificial canal in Britain," was restored to a state of perfect efficiency at a cost of about £40,000, forming an important connecting

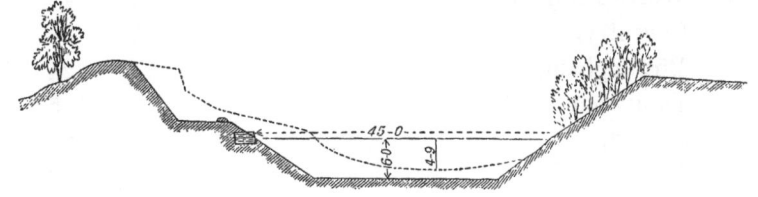

FIG. 1.

link between the Trent and Witham navigations. Fig. 1, which is a cross section at a place called "the Narrows," shows the former bed and banks in dotted lines, and the hard lines show the dimensions to which the canal was increased.

Notwithstanding the existence of this early work, however, and of some others in the country, particularly

the Sankey Brook Navigation, opened in 1760, it cannot
be doubted that the formation of the Bridgewater Canal
in Lancashire, the Act for which was obtained in 1759,
was the commencement of British Canal Navigation; and
that Francis Duke of Bridgewater, and Brindley the
engineer, who were its projectors, were the first to give
a practical impulse to a class of works which, under the
guidance mainly of Smeaton, Watt, Jessop, Nimmo,
Rennie, and Telford, has been very generally adopted
throughout the country, and has undoubtedly been of
vast importance in promoting its commercial prosperity.[1]

According to Mr. Smiles, the barge canals laid out by
Brindley, although not all executed by him, were:[2]—

	Miles.
The Duke's Canal, Longford Bridge to Runcorn,	24
Worsley to Manchester,	10
Grand Trunk, from Wilden Ferry to Preston Brook,	88
Wolverhampton,	46
Coventry,	36
Birmingham,	24
Droitwich,	5
Oxford,	82
Chesterfield,	46
	361

It is believed that the length of these inland boat-
navigations, constructed in Britain, exceeds 4700 miles.

Many of them were carried at great cost through hills
and over valleys. The Harecastle tunnel on the Grand

---

[1] *History of Inland Navigation*, particularly those of the Duke of Bridgewater, London, 1786; Hughes's Memoir of Brindley, Weale's *Quarterly Papers*,
London, 1843.

[2] Smiles's *Lives of the Engineers*.

Trunk Canal, made by Brindley, and afterwards doubled by Telford, is nearly one and a half mile in length; and the Pont-y-Cyssylte aqueduct, on the Ellesmere Canal, over the Dee, constructed by Telford at a cost of £47,000, has nineteen openings, 45 feet span, and is elevated 126 feet above the river, the canal being carried across in a cast-iron trough.[1]

It must be obvious that, to construct a navigable chan- nel through a country varying in level, and affording per- haps no great facilities for obtaining a supply of water, infers high engineering skill. Vast reservoirs must in some cases be formed for storing the water necessary to supply during dry seasons the loss by lockage, leakage, and evaporation. Feeders must be made to lead this water to the canal; hills must be pierced by tunnels; valleys must be crossed on lofty embankments, or spanned by spacious aqueducts; and, above all, the whole must be conceived and laid out with scrupulous regard to the all-important object of securing the works against injury from an overflow of water during floods, and a consequent inundation of the surrounding country. Moreover, the necessity of laying out the canal in level stretches, and surmounting elevations by means of locks or inclined planes, occurring at intervals, often occasions much difficulty, and greatly restricts the resources of the engineer. Taking, then, all these circumstances into consideration, and bearing in mind that canals were the pioneers of railways, we think it may safely be affirmed that the canal engineers of former days had much more serious *physical* difficulties

Difficulties in constructing canals.

[1] *Life of Telford,* London, 1838.

to contend with than are experienced in carrying out the
railways of modern times; if we except such works as the
Britannia Bridge, the high-level bridge of Newcastle, the
Boxhill Tunnel, and some other kindred works. But,
indeed, their *mechanical* difficulties were also greater;
for the introduction of steam, and its wide-spread application to all engineering operations, affords facilities to
the engineers of the present day which Smeaton at the
Eddystone, Stevenson at the Bell Rock, and Rennie and
Telford in their early navigation works, did not enjoy.
The distinguished merits of the engineers who practised
in the former, and at the commencement of the present
century, cannot indeed be over-estimated; and had it been
within the scope of this treatise I should readily, because
I am sure profitably, have described in detail some of the
grand aqueducts and other works on the lines of our
canals. All I can do, however, consistently with the
limits to which I have restricted myself, is to indicate the
general principles which should guide the engineer in
selecting the route for a canal, referring for details of
construction to treatises on the Strength of Materials,
Bridge-building, Tunnelling, Earthwork, Masonry, Carpen-
try, and Reservoirs, all of which are more or less applicable
to the formation of canals; and perhaps the best reference
to books on these various subjects is to be found in Professor Rankine's valuable *Manual of Civil Engineering*.

General prin-
ciples of canal
construction.

    I shall only therefore offer to the student the follow-
ing summary of engineering principles, generally applicable
to all cases.

    A canal cannot be properly worked without a full supply of water, calculated to last over the driest season of the year, and in that respect, except as to the quality of the water, demands all the care and attention requisite in investigating the sources of water for supplying towns. If there be no natural lake in the district available for supplying the canal, and affording storage for dry seasons, the engineer must select such situations as are suitable for the construction of artificial reservoirs, and the conditions to be attended to in selecting their positions are the same as those for water-works. They must command a sufficient area of drainage to supply the loss by leakage, evaporation, and lockage, due to the length of the canal, the number and size of the locks, and the probable amount of traffic. The capability of the district to afford this supply will depend on the area of the watershed, and the annual amount of rain-fall, as ascertained by accurate rain-gauge observations. The off-lets from the reservoirs must be at such an elevation as to admit of water being conveyed to the summit-level of the canal. The embankments for retaining the water must be erected on sites affording a favourable foundation, and so situated, with reference to the valley above them, that they shall, with the *minimum* height and length of embankment, dam up the *maximum* amount of water. It is further necessary to consider whether the subsoil of the valley forming the reservoirs is throughout of so retentive a nature as to prevent leakage; and it is also essential to provide, by means of waste weirs, for the discharge of floods. The Caledonian Canal, to be afterwards noticed, is, in this respect, very favourably situated, the whole supply being obtained from natural lochs. In other cases, such as the Union, Forth and Clyde, Crinan, Birmingham, and other canals, it has been found necessary to construct large artificial reservoirs, from which the water is led in feeders to points convenient for forming a junction with the canal. The water in these reservoirs is stored up in winter, and let off as required during the droughts of summer. In situations where the canal communicates with the sea or a tidal river, and where the natural supply is small, as at the Foss Dyke, already referred to, the water is raised by pumping engines.

Sectional area.   The sectional area of the barge canals constructed in this country are between 4 and 5 feet in depth. When the soil in which they were made was retentive, they

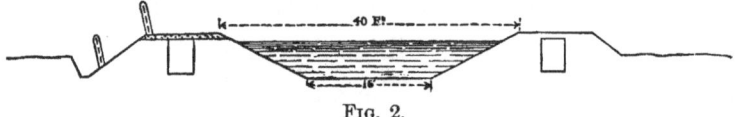

FIG. 2.

were formed as shown in the cross section, fig. 2. But when the soil was porous, clay-puddle was introduced, as shown in fig. 3. Professor Rankine says the depth of

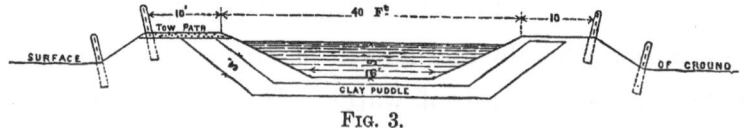

FIG. 3.

water and sectional area of water-way should be such as not to cause any material increase of the resistance to the motion of the boat beyond what it would encounter in open water, and gives the following rules as fulfilling these conditions :—

Least breadth at bottom, = 2 × greatest breadth of boat.
Least depth of water,   = 1½ foot + greatest draught of boat.
Least area of water-way, = 6 × greatest midship section of boat.

In laying out a line of canal, the engineer is much more Reaches and locks. restricted than in fixing the route of a road or railway, where gradients more or less gentle can be introduced to suit the undulating surface of the country. A canal, on the contrary, must follow rigidly the bases of sloping hills, and the windings of valleys, so as to preserve a uniform level, and it is of great importance to lay out the work in long reaches, on the same level, and to overcome elevations *in cumulo*, by means of locks at places where the nature of the country admits of its being most advantageously effected. This not only leads to a saving of attendance and expense in working the canal, but is also more convenient, as presenting fewer stoppages to the traffic; but, according to Professor Rankine, single locks are more favourable than flights of locks to economy of water. The means of overcoming the difference of level between the various reaches must depend very much on circumstances. With few exceptions, the change of level is effected by means of locks, which generally have a lift of 8 or 10 feet, though in some cases it is somewhat greater. The dimensions of the locks ought to be regulated by the traffic; but they should, in order to save water, be as nearly as possible the size of the craft to be passed through them, allowing about a foot of extra breadth and draught of water. The barge canals of which I am speaking have locks about 8 feet in breadth, and from 70 to 80 feet long, and their use in raising or lowering boats from the different reaches are so well known as not to require explanation.

The water is generally admitted into and flows from each lock by sluices formed in the gates. Sir William Cubitt, in carrying out the improvements of the Severn navigation, introduced the water through a culvert parallel to the side wall of the lock, and opening in the centre by means of a tunnel, which admits 16,000 cubic feet of water to flow into or out of the lock in $1\frac{1}{2}$ minute; and in little more than that time loaded vessels can be passed through.[1] Inclined planes and perpendicular lifts, which have the great advantage of saving water, have also been adopted so long ago as 1789 on the Ketling Canal in Shropshire, and afterwards on the Duke of Bridgewater's Canal. The most extensive application of inclined-plane navigation which I have seen, is that of the Morris Canal in the United States, constructed by Mr. Douglas of New York, on which there were no fewer than twenty-three inclined planes, having gradients of about 1 in 10, with an average lift of 58 feet. The boats weighed, when loaded, 50 tons, and after being grounded on a carriage, were raised by water-power up the inclines with great ease and expedition. The length of the Morris Canal, between the rivers Hudson and Delaware, is 101 miles, and the whole rise and fall is 1557 feet, of which 223 are overcome by locks, and the remaining 1334 by inclined planes.[2] When first describing this work in my book on American Engineer-

Inclined planes and perpendicular lifts.

---

[1] *Minutes of Proceedings of Institution of Civil Engineers,* vol. v. p. 340.
[2] Stevenson's *Sketch of Civil Engineering in North America.*

ing, I stated that the principal objection to the use of inclined planes for moving boats from different levels was
the injury they were apt to sustain in supporting great
weights while resting on the cradle during their progress
over the plane. A slimly-built canal-boat 80 feet long, and
loaded with 30 tons, could not be grounded on a smooth
surface without straining her timbers, but this objection
has to some extent been overcome on an inclined plane more Monkland
Canal incline.
recently constructed by Mr. Leslie and Mr. Bateman on the
Monkland Canal, where the boats are not wholly grounded
on the carriage, but are transported over the incline in a
caisson of boiler-plate containing 2 feet of water, and are
thus *water-borne*. This inclined plane is wrought by two
high-pressure steam-engines of 25-horse power each. The
height from surface to surface is 96 feet, and the gradient
is 1 in 10. The maximum weight raised is from 70 to 80
tons, and the whole transit is accomplished in about ten
minutes. For the five years previous to the end of 1856,
the average number of boats that passed over the incline
each year was 7500. Mr. Green introduced on the Great
Western Canal a perpendicular lift of 46 feet. Sir William Cubitt has also introduced three inclined planes,
having gradients of 1 in 8, on the Chard Canal, Somerset-
shire. One of these inclines overcomes a rise of 86 feet ;
and they are said to act very satisfactorily.[1]

An essential adjunct to a canal is a sufficient num- Waste weirs.
ber of waste weirs to admit of the discharge of the surplus water which accumulates during floods, and which
may, if not provided with an exit, rise to such a height as

[1] *Minutes of Proceedings of Institution of Civil Engineers*, vol. xiii. p. 205.

to overflow the towpath, and cause a breach in the banks, producing stoppage of the traffic and damage to the adjoining lands. In determining the number and positions of these waste weirs, the engineer must be guided entirely by the nature of the country through which the canal passes. Whenever an opportunity occurs of discharging surplus water from an aqueduct into a stream crossed by the canal, a waste weir may safely be introduced; but, independently of this natural facility, the engineer must consider from what quarters, and at what points, the greatest influx of water may be apprehended, and must at such places not only form waste weirs of sufficient size to let off the surplus, but prepare artificial courses for its discharge into the nearest streams. These waste weirs are overflows placed at the top water-level of the canal, so that in the event of a flood occurring, the water flows over them, and thus relieves the banks. The want of sufficient escape for flood-water has occasioned overflows of canal banks which were attended with very serious injury to the works, and lengthened suspension of the traffic; and attention to this particular part of canal construction is of essential importance.

Stop-gates. It is necessary to introduce stop-gates at short intervals of a few miles, for the purpose of dividing the canal into isolated reaches, so that in the event of a breach occurring, the gates may be shut, and the discharge of water confined to the small reach intercepted between two of them, instead of extending throughout the whole line of canal. In large works these stop-gates may be most advantageously formed in the same manner as the gates of locks, two pairs of gates being made to shut in opposite directions. In small works they may be made of thick planks, which are slipped into grooves formed at those narrow parts of the canal which occur under road bridges, or at contractions made at intermediate points to receive them. Self-acting stop-gates have been tried, but their success has not been such as to lead to their general introduction. Stop-gates are further found to be very useful in cases of repairs, as they admit of the water being run off from a short reach, when the repairs can be made, and the water afterwards restored, with comparatively little interruption to the traffic. Their value in obviating serious accidents was well exemplified on one occasion in my own experience. The water during a heavy flood flowed over the towing-path of the Union Canal connecting Edinburgh and Glasgow, near the end of an aqueduct which adjoined a high embankment, and the uncontrolled current carried away the embankment and the soil on which it rested, to the depth of eighty feet, as measured from the top water-level. The stop-gates were, on the occasion referred to, promptly applied, and the discharge confined to a short reach of a few miles, otherwise the injury (which was, even in its modified form, very considerable) would have been enormous, not only to the canal works, but to the adjoining lands.

For the purpose of draining off the water to admit Off-lets. of repairs after the stop-gates have been closed, it is proper to introduce, at convenient situations, a series of exits called " off-lets," which are pipes placed at the level of the bottom of the canal, and fitted with sluices which can be opened and shut when required. These off-lets are generally formed at aqueducts or bridges crossing rivers where the contents of the canal can be run off into the bed of the stream, the stop-gates on either side being closed, so as to isolate the part of the canal from which the water is withdrawn.

Towpaths. In executing the work, provision must be made for the proper drainage of the towpath, which should be made highest at the side next the canal, and sloped with **Drainage and** a gentle inclination towards the outside. The drainage **Pitching.** of the towpath should be carried to a sky drain, and at intervals passed below it into the canal, as shown in fig. 4.

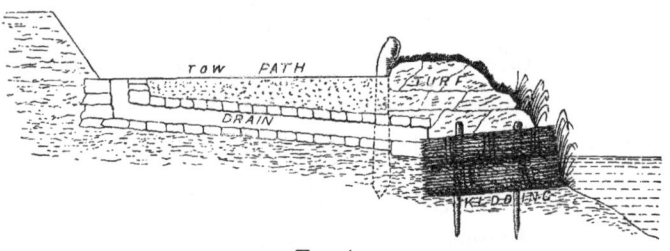

FIG. 4.

The preservation of the banks at the water-line is also a matter of importance. "Pitching" with stones and "facing" with brushwood are employed, and, in my experience, the latter, if well executed, forms an economical and effectual protection. Fig. 4 is a section of these works as executed at the Foss Dyke.

In forming the *alveus* or bed of the canal, great care must be taken, especially on embankments, and even in cuttings, where the soil is porous, to provide against **Puddling.** leakage by using puddle, as shown on fig. 3, page 12. And here it is proper to remark, that an all-important matter, as affecting the construction of the works, is the possibility of getting clay in the district, or such other soil as may be worked into puddle, on the good quality of which the stability of the reservoir embankments, and the imperviousness of the beds and banks of the canal, mainly depend.

These are the only points of general application, in the construction of canals, to which I can advantageously refer; and in applying them to each case, the student must be guided, *first*, by theoretical knowledge, to be acquired by a careful study of his profession; and, *secondly*, by that experience which can be gained only by attendance on works in actual operation.

The best mode of conducting traffic on canals and rivers is hardly within the limits of this treatise, seeing that it is a subject not directly connected with the construction of canals or the improvement of rivers, but rather with the use made of them after having been constructed or improved. Mode of conducting traffic on canals.

Not a little, however, has been written on the subject, and I refer the reader who wishes to study it fully, to the observations made by Mr. Walker and Mr. George Rennie, in the *Transactions of the Royal Society* and of the *Institution of Civil Engineers*, and especially to the very valuable researches on hydrodynamics by Mr. J. Scott Russell, in the *Transactions of the Royal Society of Edinburgh*.

These investigations were made before the general establishment of railways, at a time when swift canal travelling seemed still a desirable attainment, and nothing can be more interesting than Mr. Russell's able treatment of that subject.　But though *boats* propelled at high speed on canals have given place to *railway carriages*, yet the canal traffic must be conducted, and the cheapest means of effecting the "haulage" with the least danger to the banks is still an important inquiry, and has within the last few years afforded matter for some highly interesting papers and statements in the *Proceedings of the Institution of Civil Engineers.*　These are communications on the employment of steam power on the Gloucester and Berkeley Canal, by Mr. W. B. Clegram ; on the Grand Canal, Ireland, by Mr. Healy ; on the Forth and Clyde, by Mr. J. Milne ; and on the Aire and Calder, by Mr. W. H. Bartholomew, to all of which I refer as containing valuable information.[1]

Wasting of the banks.

One of the great objections to high speeds on narrow channels is the wasting of the banks by the displacement produced in propelling the vessel through the water.　The wasting indeed takes place even with very low speeds, and as a matter of canal engineering it is necessary to notice it. To give an instance of the effect on the large scale: Mr. Ure says that the river-steamers on the Clyde, going at a speed of eight to nine miles per hour, produce a swell which commences to rise when the vessel is "two or three miles off" —a circumstance which was first noticed by Mr. J. Scott Russell in 1837.　The swell gradually increases as the steamer approaches, and at last, becoming a wave of translation, it breaks on the river walls nearly abreast of the vessel, following her on her course along the river, as a violent breaking wave, measuring sometimes 8 or 10 feet from the

---

[1] *Minutes of Proceedings of Institution of Civil Engineers,* vol. xxvi. p. 1.

hollow in the channel to the crest on the wall. A coating of heavy whinstone rock, from 2 to 3 feet thick, extending from low to high water mark, is found necessary to enable the banks to withstand it. Mr. Ure also found that the action of passing steamers, though very destructive to the banks, was useful in stirring up the mud from the bottom, which was carried off by the currents to an extent which he estimates to be from 20 to 25 per cent. of the whole quantity dredged from one particular part of the river where he carefully measured it. It will at once be apparent, that however inconvenient these wasting waves may be in a river, the waves in a canal, though smaller, are nevertheless a source of greater anxiety, acting as they do in a narrow artificial channel formed at some places on high embankments, the failure of which might be attended with serious consequences.

The wasting on canals where the traffic is conducted at a moderate speed is found to extend not more than 18 inches to 2 feet, that is, 1 foot above and below the water-line, and Mr. Clegram states that he has found on the Gloucester Canal that a facing of stone filled into a recess cut in the banks formed a complete protection. The stone facing is about a foot in thickness, and is formed of stones from 18 to 20 cubic feet. The protection adopted at the Foss Dyke Canal consisted of fascines of brushwood, as shown on page 18, and was found to be most effective.

What has lately led to the consideration of the best means of protecting the banks of canals is the substitution of steam for horse power in working the traffic, and

Steam-towing on Gloucester and other canals.

this has been entirely successful. The first attempt at using steam-power on canals was made on the Forth and Clyde Canal with Symington's boat in 1789. Mr. Milne states the results of various experiments to introduce tugs, but these were ultimately abandoned in favour of steam-lighters, which now in great numbers navigate the canal, and make passages to Leith, Greenock, and other trading ports on the Firths of Forth and Clyde.

This system, however, would not suit the trade of the Gloucester Canal, which is chiefly frequented by sea-borne vessels, and Mr. Clegram, its engineer, gives an interesting account of the introduction of steam-towing on that navigation. The following extracts from his paper seem generally applicable to all navigations where towing is to be adopted. He says the navigation is a ship canal leading from the Severn at Gloucester to the Severn at Sharpness Point. It is 16½ miles in length, and has a depth of water varying, according to the season, from 18 feet to 18 feet 6 inches. Vessels up to 600 tons and 700 tons register navigate the canal to Gloucester. Prior to the year 1860 all sea-going vessels passing through were towed by horses, the number of horses being regulated by a scale varying from 1 horse for a vessel of 40 tons to 9 horses for a vessel of 420 tons. The cost of this amounted generally to about one farthing per ton per mile on the register tonnage of the vessel. The speed varied from 1 mile to 3 miles per hour, according to the size of the vessel and the state of the weather.

In 1860 steam-tugs were placed upon the canal to do this work. They are iron boats, 65 feet long, 12 feet beam, and draw 6 feet 3 inches of water, fitted with high-pressure engines, the diameters of the cylinders are 20 inches, with a length of stroke of 18 inches, the pressure of the steam being 32 lb. on the inch, and the cost of each £3000. Nearly the whole of the sea-going craft are now towed by these tugs. The vessels range from 30 tons up to 600 and 700 tons register, with a varying draught of water of from 6 to 16 feet. They are towed either singly or in a train, according to circumstances. Sometimes as many as nine, ten, and even thirteen loaded vessels of from 50 to 100 tons register have been towed by one tug at the rate of 3 miles to $3\frac{1}{2}$ miles an hour. The heaviest load drawn by any one tug has been 1690 tons of goods, in three vessels. Their draught of water varied from 14 feet 6 inches to 15 feet 6 inches, and they were taken the whole length of the canal at the speed of 2 miles an hour. The smaller vessels are towed at a speed of 4 miles an hour, to which as a rule they are restricted.

The employment of steam as a towing power has been found in nearly every way advantageous. The work is greatly economized. The vessels rub much less against the sides of the banks, the power being right ahead, and not on one side, as with horses. The wear on the ropes used in tracking is much reduced, the speed is increased, and vessels can now be moved along the canal in weather which would have prevented horses doing the work. With a strong wind athwart the canal vessels cannot be tracked in train; they must then be taken singly, or at most two at a time. When vessels are towed in train, as a rule the largest and heaviest draughted are placed first, and the hawser leading from the first vessel to the tug, is taken from each side of the bow. With this arrangement, and a skilful management of the tug, the vessel can be kept fairly in the line of the canal.

The one and only disadvantage of this system, on a canal the sides of which are unprotected, is the additional wear caused by the constant passage of the tugs as well as by the run of water between the sides of the large vessels and the banks. Such vessels occupy a large part of the sectional area of the canal, and being taken along at a much greater speed than they were by horses, the back run of water is more rapid and prejudicial. When the vessels or trains of vessels are heavy, and the tugs are working up to their full power and speed, the water thrown back by the action of the screw against the bow of the first vessel is thrown off by it to the banks on either side, and is the cause of considerable wash. This has been attempted to be remedied by placing the first vessel farther back from the tug ; but in practice it is found that a distance of from 40 to 50 feet is the farthest separation that can be allowed without sacrificing that *hold* between the two which prevents the vessel sheering from side to side. The first vessel, being kept steadily in her course, the others follow without much difficulty.

The employment of tugs has afforded an unexpected facility in cleansing the canal from the deposit of mud. Formerly it was difficult to remove this deposit from the slopes of the banks. It was dangerous to apply the dredger, and although the mud in the bottom of the canal could be removed, it collected on the slopes, and at times inconveniently contracted the capacity of the canal. Since the vessels have been moved at greater speed and in trains, this deposit has been entirely removed from the slopes to the bottom of the canal, whence it can readily be taken out by the dredger.

Steam-power is even more important as connected with the traffic on navigable rivers, but as already stated, I do not propose to enter upon it, but must refer to treatises on Steam Navigation; I shall only remark that while it is conducted on our narrow rivers by employing, as in the case of the Tees, sometimes as many as three tugs to take a large iron steamer of 3000 tons from the shipbuilding yards of Stockton to the sea, in America the process is reversed, one large powerful steamer being employed on the capacious rivers to tow a whole fleet of vessels. The towing of vessels on the Mississippi and St. Lawrence has been brought to great perfection. I had an opportunity of witnessing this on the St. Lawrence; having passed from Quebec to Montreal in a large powerful tug-steamer, carrying goods and many hundreds of passengers, and having no fewer than five sea-borne vessels in tow, drawing from 7 to 12 feet of water. These vessels were all towed by separate warps, and were ranged astern of each other in two lines, three of them being made fast to the larboard and two to the starboard side of the vessel. The management of a large steamer with so many vessels *in tow*, in the intricate navigation and strong currents of the St. Lawrence, required no small amount of skill; but when it was necessary to stop the

Steam-towing on rivers.

steamer to take in fuel, the captain dropped the vessels astern and again picked them up on resuming his course, with a dexterity which I have never seen equalled, and we made the passage of 180 miles in forty hours, being at the rate of $4\frac{1}{2}$ miles an hour, against a current averaging 3 miles per hour.

CHAPTER II.

SHIP CANALS.

Utility of Ship canals—Languedoc, Forth and Clyde, and Crinan Canals—Ship canals divided into three sections: those through high districts of country; through low-lying districts; and those without locks, deriving their water-supply from the sea—Caledonian Canal—Canals of .North Holland—Amsterdam Canal—Suez Canal.

THE statement at the beginning of the last chapter, as to railways having superseded canals, applies to the smaller canals we have been considering, but is not true of the larger class of works still to be noticed. Ship canals undisturbed by competing schemes retain all their useful-ness, and indeed in the recent construction of the Suez and new Amsterdam Canals have acquired an importance before unclaimed for works of that class. Their use-fulness in affording a short and sheltered passage for sea-borne vessels, enabling them to escape tedious and sometimes dangerous coasting voyages, was early ac-knowledged, and can hardly be over-estimated.

The Languedoc Canal, already mentioned, by a short passage of 148 miles saves a sea voyage of upwards of 2000 miles through the Straits of Gibraltar. The Forth and Clyde Canal, projected by Smeaton in 1764, and opened in 1790, enables sea-borne vessels, not exceeding 8½ feet draught of water, to pass from opposite coasts of Scotland by 35 miles of inland navigation; and the

Crinan Canal substitutes a short inland route of 9 miles for a sea voyage round the Mull of Kintyre of about 70 miles.

To most of the early ship canals that have been executed, the principles of construction stated in the preceding chapter are generally applicable—the depth of water and the dimensions of the locks being increased to admit the larger size of craft which use them,—and therefore I do not propose to describe them further; but it would not do to dismiss the subject without referring in detail to some of the largest of these canals, in order to illustrate the different character of work employed to suit the varied physical aspects of the countries through which they pass, and I think the student of engineering will find that the works themselves are so interesting as to demand special notice.

Ship canals divided into three classes.

I propose to divide ship canals into three sections :—

First, Canals which on their route from sea to sea traverse high districts, surmounting the elevation by locks supplied by natural lakes or artificial reservoirs, such as Languedoc or Caledonian Canals.

Second, Canals in low-lying districts, which are carried on a uniform water-level from end to end, and are defended against the inroad of the sea at high water by double-acting locks, which also retain the canal-water at low tide, such as the canals of Holland and other low-lying countries.

Third, Canals of which the Suez is the only example yet made, without locks at either end, and communicating freely with the sea, from which it derives its water-supply.

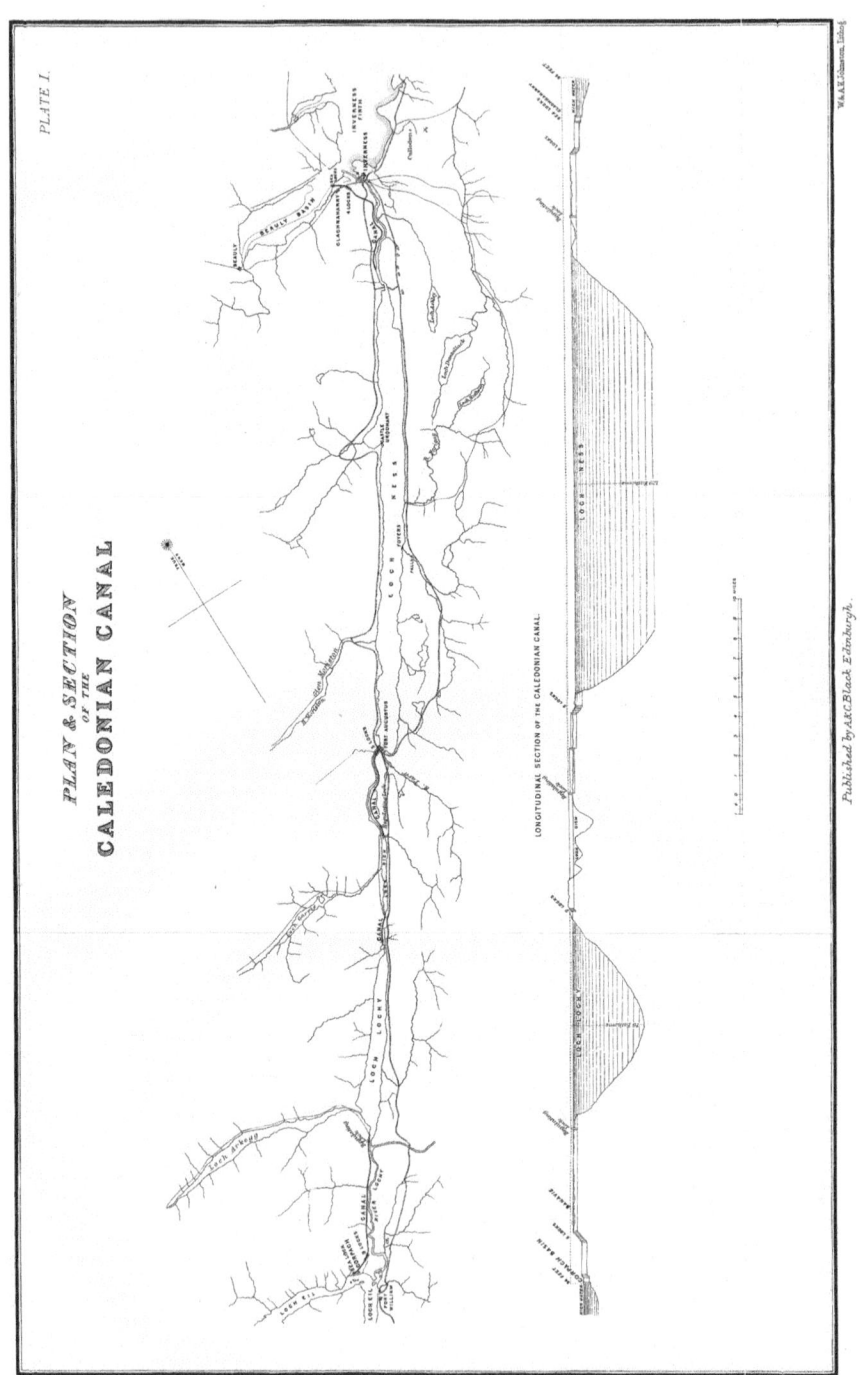

PLATE I.

PLAN & SECTION
of the
CALEDONIAN CANAL

LONGITUDINAL SECTION OF THE CALEDONIAN CANAL.

Published by A&C.Black Edinburgh.

The material originally positioned here is too large for reproduction in this reissue. A PDF can be downloaded from the web address given on page iv of this book, by clicking on 'Resources Available'.

CALEDONIAN CANAL.

Of each of these three sections I shall give an example. And I believe that the Caledonian Canal is as good a specimen of the *first* class as can be selected.

So early as 1773, James Watt was employed to survey the country between the Beauly at Inverness and Loch Eil, at the mouth of the river Lochy—a distance of about 60 miles,—with the view to the formation of a ship canal between the two seas, to save about 400 miles of coasting voyage by the north of Scotland, through the stormy Pentland Firth. The district referred to, called the "Great Caledonian Glen," as will be seen from Plate I., embraces a chain of fresh-water lakes, which, in connexion with the surrounding glens, have afforded an interesting field for the speculations of the geologist; and no doubt the first conception of a canal through the district owed its origin to the apparent facilities for inland navigation which the lakes afforded.[1] In 1801 Telford was employed by Government to report, and the ultimate result of that report was the construction of the canal, which was opened in 1823.

The summit-level of the canal is at Laggan, which is situated between Loch Oich and Loch Lochy, and from this place the drainage flows to the eastern and western seas. The district which discharges into the eastern outlet comprehends an area of about 700 square miles, chiefly of high mountainous country, intersected by streams and lakes, which discharge themselves into Loch Oich, Loch

[1] *Life of Telford :* Caledonian Canal.

Ness, and Loch Doughfour, and thence are conveyed into the Moray Firth by the river Ness.

Loch Oich, the summit-level of the canal, has an area of about 2 square miles, and the present standard level of its surface is understood to be 102 feet above the level of mean high-water of neap tides in Beauly Firth. It receives the drainage of Loch Quoich and Loch Garry. The waters of Loch Oich are discharged through the river Oich into Loch Ness, which is about 24 miles in length, and has an area of about 30 square miles. Loch Ness receives the waters of the Tarff, the Foyers, and Glenmoriston, and the drainage of numerous other streams and lakes of less note. It discharges its waters through a comparatively narrow neck, called Bona Passage, into the small loch of Doughfour, from whence they find an exit to the Beauly and Moray Firths by the river Ness, on which the town and harbour of Inverness are situated. The drainage of the western district of the country, including Loch Arkegg, finds its way into Loch Lochy, which is about 10 miles long, and thence, by the river Lochy, to the western sea at Loch Eil.

The two locks in Loch Beauly, at the northern entrance to the canal, are each 170 feet long, 40 feet wide, and have a lift of about 8 feet. At Muirtown, a little farther on, are four locks, of 180 feet in length and 40 feet in width, having a rise of 32 feet, raising the canal to the level of Loch Ness, which it enters at Bona. The works westward of Loch Ness consist of an artificial canal, with seven locks, communicating with Loch Oich. Between Lochs Oich and Lochy are two locks. At the south end of Loch Lochy is a regulating lock, and the canal is carried from this point on the level of Loch Lochy to Bannavie, where it descends 64 feet, by eight connected locks, forming what is called in the country "Neptune's Staircase;" finally, at Corpach, the canal descends, by two locks, to the level of Loch Eil.

Of the whole distance, about 37½ miles may be taken as natural lake navigation, and the remaining 23 as artificial.

The artificial canals were made 120 feet in width at top water level, 50 feet at bottom, and 20 feet in depth. In the course of inquiries as to the state of the canal in 1849, under a remit from the Admiralty, I found that the shallows at Loch Oich, and the cutting at the summit-level, had not been carried to the full depth, and an additional depth had been gained at that place by raising the level of Loch Oich; but still I was led to the conclusion that the standard depth of the canal cannot be regarded as more than 18 feet, giving access to vessels of 160 feet in length, 38 feet beam, and 17 feet draught of water.[1]

In carrying out this remarkable work Telford had to deal with difficulties of no ordinary kind, in rendering available rugged Highland lakes, and surmounting the summit-level of the glen. The work, which cost about one million sterling, is a noble monument of his engineering skill.

The canals of Holland are specimens of the *second* Canals of North Holland. section of works to which I referred, and of these a

[1] Report on the Caledonian Canal to the Admiralty, 1849, by James Vetch, R.E., and David Stevenson, C.E.

very remarkable one is the North Holland Canal, designed by M. Blanken, and completed in 1825, who, instead of the Highland glens of Scotland, had to deal with the proverbial *lowness* of the country, and to protect his works from the assaults and encroachments of the waves, and when I examined the work they were locking vessels *down* from the sea into the canal. It extends from Amsterdam to the Helder, is 50 miles in length, and is formed of the cross section shown in fig. 5. It enables vessels trading from Amsterdam to avoid the islands and sandbanks of the dangerous Zuider Zee, the passage

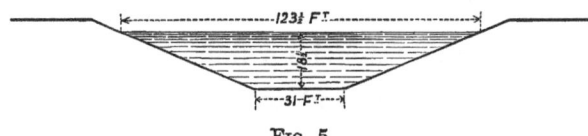

FIG. 5.

through which, in former times, often occupied as many weeks as the transit through the canal now occupies hours.

It was here that Bakker, a burgomaster of Amsterdam, in 1688, introduced his "Camel" for floating large vessels over the shoals of the Pampus between Amsterdam and the Texel Roads, by means of which, according to Sir John Leslie, an Indiaman which drew 15 feet water had its draught reduced to 11 feet. The following description is given of these camels by Mr. G. B. W. Jackson: [1]—"These water camels, of which several were kept at Pampus on the Y stream, were used in pairs whenever they were required. They were of sufficient length to suit the largest vessel, being each pro-

[1] *Minutes of Proceedings of Institution of Civil Engineers*, vol. vi. p. 82.

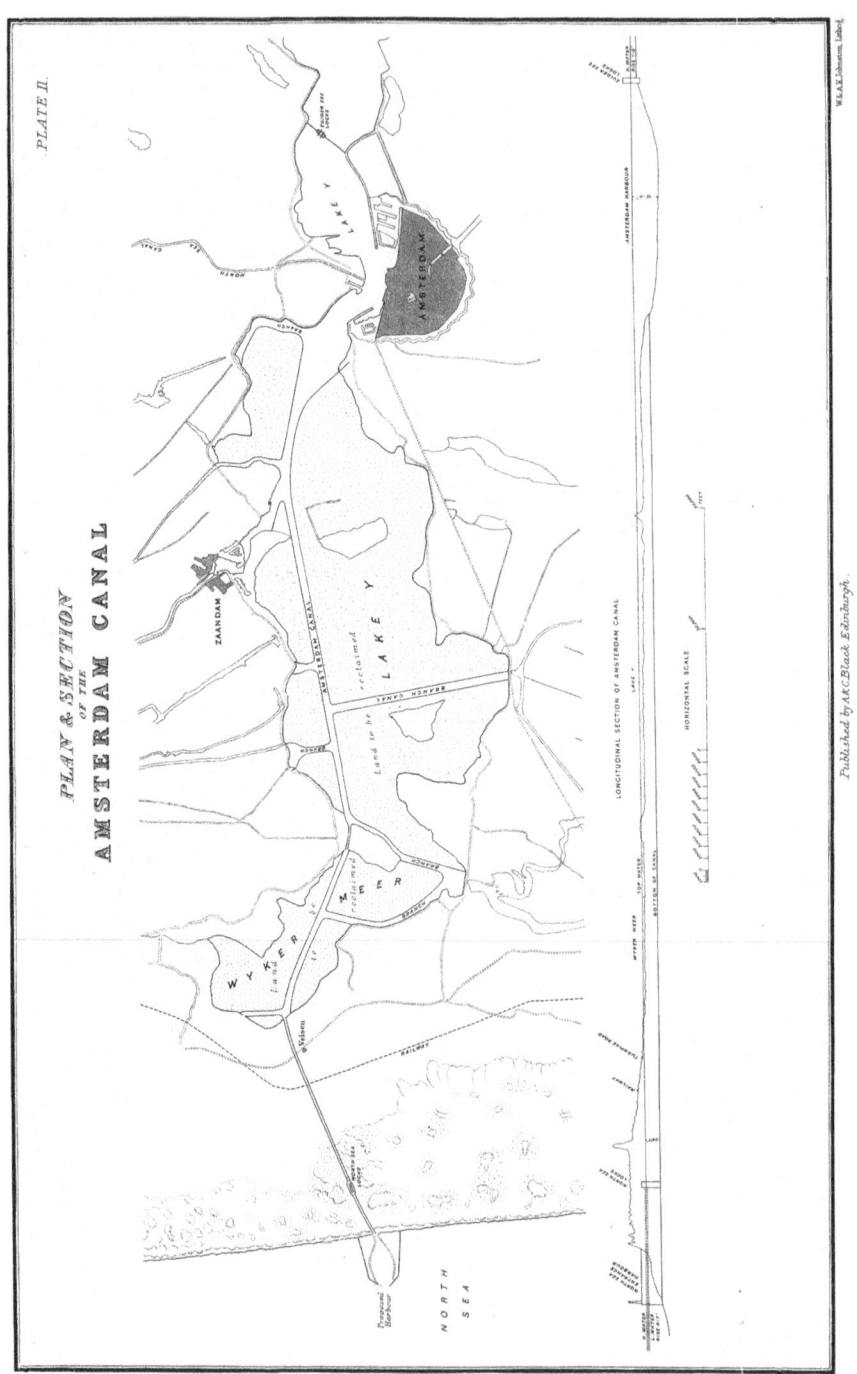

PLATE II.

PLAN & SECTION OF THE AMSTERDAM CANAL

Published by A&C.Black Edinburgh.

The material originally positioned here is too large for reproduction in this reissue. A PDF can be downloaded from the web address given on page iv of this book, by clicking on 'Resources Available'.

vided with a rudder, and with windlasses on the outer side, to which the ropes were attached for securing the vessel. They probably had 31 ship's pumps of 6 inches diameter for clearing, and about 16 valves for letting in the water used to sink them. The complement of men for working them was about 50 to each. They weighed about 450 tons, drew about 2 feet 3 inches when empty, and if weighted with 820 tons about 7 feet 5 inches more. They have been broken up, as being no longer required.

AMSTERDAM CANAL.

But the North Holland Canal, which has long proved so useful to the commerce of the district, is destined soon to be superseded by the New Amsterdam Canal, a work of great magnitude, which I propose to describe as an illustration of ship canals of the *second* class, having received, through the kindness of Mr. J. C. Hawkshaw, the following interesting details regarding it :—

The rapid increase in the trade of the ports to the southward and eastward of the Helder, effected by the construction of railways throughout Europe, rendered it imperative for the merchants of Amsterdam to provide better communication with the North Sea than that afforded by the North Holland Ship Canal, already noticed, or suffer its trade to pass to other ports more favourably situated for over-sea trade.

In 1865 a company was accordingly formed for the purpose of constructing a canal from Amsterdam, in nearly a direct line, to the North Sea, through Lake Y and Wyker Meer, a distance of 16½ miles. Mr. Hawkshaw and Mr. Dirks were appointed the engineers to carry out the work, a plan and section of which is given in Plate II.

The harbour in which the canal terminates in the North Sea is formed by two piers built of concrete blocks founded on a deposit of rough basalt. The piers are each 5069 feet in length, and enclose an area of about 260 acres. About 140 acres of this area is to be dredged to a depth of $26\frac{1}{4}$ feet, the remainder is to be left at the present depth for the accommodation of small craft and fishing-boats.

From its commencement at the harbour the canal passes by a deep cutting through a broad belt of sand-

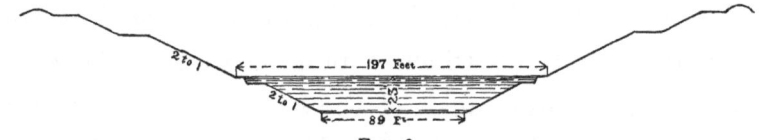

FIG. 6.

hills which protect the whole of this part of the coast of Holland from the inroads of the sea. The cross section of the canal at this place is shown, fig. 6. This cutting is about 3 miles in length; the greatest depth from the surface to the bottom of the canal is 78 feet; and the amount of earthwork excavated is 6,213,000 cubic yards. On emerging from the sand-hills the canal passes by the village of Velsen, in the neighbourhood of which it is crossed by the railway from Haarlem to the Helder, and there enters the Wyker Meer, a large tract of tide-covered land. After traversing the Wyker Meer it passes by a cutting of 327,000 cubic yards through the promontory called Buiten-hinsin, which separates that Meer from

Lake Y, another large tide-covered area.   The rest of its
course lies through Lake Y as far as Amsterdam.

There are two sets of locks, one set at either end.
The North Sea locks are at a distance of about three quarters of a mile from the North Sea harbour.   These locks,
as shown in fig. 7, consist of three passages.   The central
or main one is 60 feet wide and 390 feet long, and will be

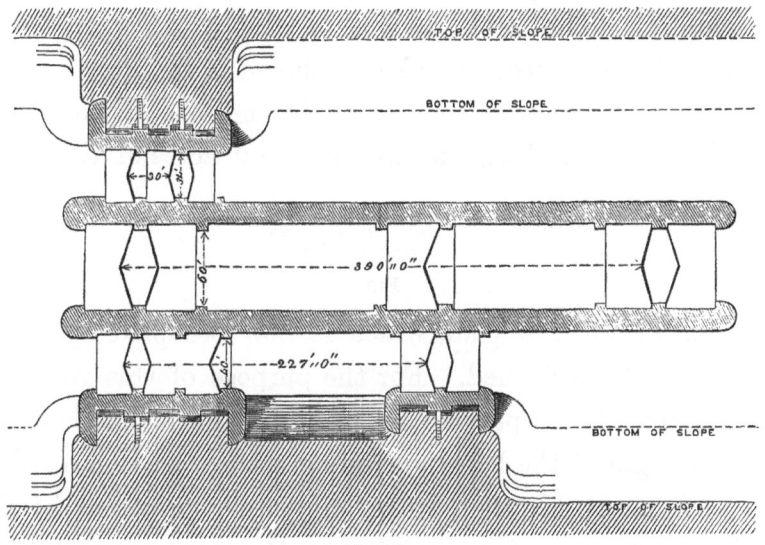

FIG. 7.

furnished with two pairs of gates at each end, pointing
in opposite directions, and one pair in the centre.   The
northernmost side passage for barges is 30 feet long and 34
feet wide, with three pairs of gates; that to the south is 227
feet in length and 40 feet wide, with five pairs of gates.

In constructing the canal, which is now far advanced
towards completion, the cuttings were first begun.   The
material proceeding from these cuttings was conveyed
either by means of waggons or barges, and deposited so

as to form two banks 443 feet apart, through the lakes
on each side of the main canal, as shown by the hard
lines on the plan, and also to form the banks of the branch
canals on either side.   The total length of these banks is
38½ miles.   The nucleus of the bank is formed of sand,
with a coating of clay, and protected during its progress
with fascines; and when the banks are far enough advanced, the deep channel for the canal is excavated by
dredging.   The cross-section of the canal and banks
through these meers or lakes is shown in fig. 8.

The formation of the banks through the Wyker Meer
and Lake Y will enable about 12,000 acres of the area,

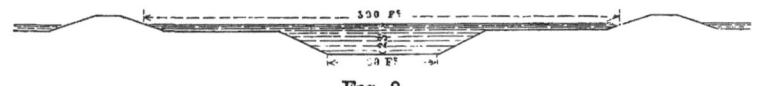

FIG. 8.

as shown on the plan, which is now occupied by these
lakes, to be reclaimed.   For the purpose of this reclamation, and also to provide for the drainage of the land on
the margin of the lakes, including a large portion of what
was formerly Haarlem Meer, pumps are provided by the
Company at various points on the main and branch
canals.   The Canal Company are bound to keep the surface-water of the canal about 1 foot 7 inches *below* average
high-water level.   In order to insure this level being
maintained, three large pumps have been erected in connexion with the locks hereafter to be described, on the
dam between Amsterdam and the Zuider Zee.   They consist of three Appold pumps, the largest of the kind yet
made, the fans being 8 feet in diameter.   Each pump is
worked by a separate engine of 90 nominal horse-power.

The maximum lift is 9 feet 9 inches, at which the three pumps are capable of discharging 1950 tons a minute; with the ordinary working lift of 3½ feet they will discharge 2700 tons a minute. Pumps of a similar construction, with fans one foot less in diameter, erected some years ago under Mr. Hawkshaw's direction, at Lade Bank, in Lincolnshire, have been found to work satisfactorily.

Lake Y, as will be seen from the plan, extends about 4½ miles to the eastward of Amsterdam; and here it was necessary to form a dam with locks for the passage of vessels. The dam crosses Lake Y at a point about two miles to the eastward of Amsterdam, where it is contracted to 4265 feet in width.

As it was necessary to construct these locks before completing the dam across Lake Y, a circular cofferdam, 590 feet in diameter, consisting of two rows of piles, 49 feet long, was constructed in the tideway, and within this dam the locks were built.

These locks have three main passages, each with five pairs of gates, and one smaller passage with three pairs of gates, arranged much in the same manner as the North Sea locks, shown in page 35, but their dimensions are not so large. The central main passage has a length of 315 feet, and is 60 feet wide. The passages on each side of it have each a length of 238 feet, and are 47 feet wide. The small passage is 30 feet long and 34 feet wide. There are also three sluiceways for the pumps, each 110 feet long and 13 feet wide, and each provided with three pairs of gates.

The whole of the masonry and brickwork for these

locks and sluiceways was founded on bearing-piles, upwards of 10,000 in number.

The bottom where the cofferdam was placed, consisted
of mud, and some difficulty was experienced in maintaining it till the work was completed.

The dam across Lake Y, as shown in section, fig. 9,
consists of clay and sand, placed on and protected at the
sides by large masses of wicker-work, which is afterwards

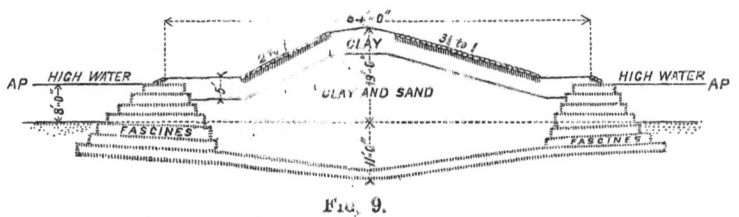

Fig. 9.

covered with basalt in the manner usually adopted in
Holland.

All the lock gates at both ends of the canal pointing
seawards are of malleable iron; the gates pointing inwards towards the canal are of wood.

The necessity, for drainage purposes, of maintaining
the surface-water of the canal at the prescribed low level
calls for a sufficient barrier being provided against the sea
at both ends, as the sea level will not unfrequently, at
high water, be several feet above the level of the canal.

This necessity, as well as the difference of level and
periods of high water in the Zuider Zee and the North
Sea, required a totally different design from the Suez
Canal, to be afterwards described.

The contract sum for the execution of the Amsterdam Canal is 27,000,000 of florins, or about £2,250,000.

It is to be hoped that ships may pass through the canal in about two years.

SUEZ CANAL.

Of the *third* section of works there exists as yet only a single example, in the Suez Canal, one of the most remarkable of the engineering works of modern times ; but though it is called a canal, it bears little resemblance to the works I have been describing under that name, for it has neither locks, gates, reservoirs, or pumping-engines, nor has it, indeed, anything in common with canals, except that it affords a short route for sea-borne ships. It is in fact, correctly speaking, an artificial arm of the sea, or strait connecting the Mediterranean and the Red Sea, from which it derives its water-supply.

The idea of forming this connecting link is of very ancient origin, and its author is unknown. It is understood, however, that a water communication between the two seas, for small vessels, was formed as early as 600 years before the Christian era, and existed up to a period of about 800 years after that date, and then was allowed to fall into disuse. The idea of restoring this ancient communication on a scale suited to modern times is understood to be due to Napoleon I., who, about the close of the last century, obtained a report from M. Lepère, a French engineer, which however was followed by no result, and it remained for M. de Lesseps, in the present day, to realize what were thought the dreams of commercial speculators, by carrying out the long-desired passage between the two seas. But the postponement of the

scheme unquestionably favoured the chances of its commercial success, for had the canal been completed even a
few years earlier, comparatively few vessels would have
been found to take advantage of it. Sailing vessels would
never have navigated the Mediterranean and encountered
the passage through the canal, and the tedious and difficult voyage of the Red Sea. They would undoubtedly
have preferred to round the free seaway of the Cape of
Good Hope, with all its ocean dangers and excitements,
to threading their way through such an inland passage,
involving risks of rocks and shoals, protracted calms
and contrary winds. But the introduction of ocean-
going screw-steamers was an entirely new feature in
navigation. Being independent of wind for their pro-
pulsion, and admirably fitted for navigating narrow passages and seas, their rapid and general adoption by all
the leading shipping firms in the country afforded not
only a plea, but a necessity for the short communication
by the Mediterranean and Red Sea. It was indeed
a great achievement to reduce the distance between
Western Europe and India from 11,650 to 6515 miles,
equal, according to Admiral Richards and Colonel Clarke,
R.E., to a saving of 36 days on the voyage; and this is
the great result effected by cutting the Suez Canal between the Mediterranean and the Red Sea.

Mr. Bateman, C.E., who visited the canal as the
representative of the Royal Society, communicated to
that body an interesting description of the works, in
which he gives the following account of the early proposals and negotiations of M. Ferdinand Lesseps, who

has the credit of having brought the work to a successful issue :[1]—

" The project " of M. Ferdinand Lesseps " was to cut a great canal on the level of the two seas, by the nearest and most practicable route, which lay along the valley or depression containing Lake Manzaleh, Lake Ballah, Lake Timsah, and the Bitter Lakes. The character of this route was well described in 1830 by General (then Captain) Chesney, R.A., who examined and drew up a report on the country between the Mediterranean and the Red Sea. At that time a difference of 30 feet between the two seas was still assumed, and all proposals for canals were laid out on that assumption. Allowance must, of course, be made for this error in so far as it affected any particular project of canal ; but it would not affect the accuracy of any general description of the district to be traversed. General Chesney summed up his report by stating, ' As to the executive part there is but one opinion : there are no serious difficulties ; not a single mountain intervenes, scarcely what deserves to be called a hillock ; and in a country where labour can be had without limit, and at a rate infinitely below that of any other part of the world, the expense would be a moderate one for a single nation, and scarcely worth dividing among the great kingdoms of Europe, who would all be benefited by the measure.'

" M. Lesseps was well advised therefore in the route he selected, and (assuming the possibility of keeping open the canal) in the character of the project he proposed.

[1] *Proceedings of the Royal Society*, 1870, p. 132.

"From 1849 to 1854 he was occupied in maturing his project for a direct canalization of the isthmus. In the latter year Mahomet Saïd Pasha became Viceroy of Egypt, and sent at once for M. Lesseps to consider with him the propriety of carrying out the work he had in view. The result of this interview was, that on the 30th of November in the same year a commission was signed at Cairo, charging M. Lesseps to constitute and direct a company named 'The Universal Suez Canal Company.' In the following year, 1855, M. Lesseps, acting for the Viceroy, invited a number of gentlemen, eminent as directors of public works, as engineers, and distinguished in other ways, to form an International Commission for the purpose of considering and reporting on the practicability of forming a ship canal between the Mediterranean and the Red Sea. This Commission, which included some of the ablest civil and military engineers of Europe, was honorary, and its members were considered as guests of the Viceroy.

"The Commission met in Egypt in December 1855 and January 1856, and, accompanied by M. Lesseps and by Mougel Bey and Linant Bey, engineers, and other gentlemen in the service of the Viceroy, they made a careful examination of the harbours in the two seas, and of the intervening desert, and arrived at the conclusion that a ship canal was practicable between the Gulf of Pelusium in the Mediterranean and the Red Sea near Suez. They differed, however, as to the mode in which such a canal should be constructed. The three English engineering members of the Commission were of opinion that a ship canal having its surface raised 25 feet *above* the sea-level, and communicating with the Bay of Pelusium at one end and the Red Sea at the other, by means of locks, and supplied with water from the Nile, was the best mode of construction. The foreign members, on the contrary, held that a canal having its bottom 27 feet *below* sea-level, from sea to sea, without any lock, and with harbours at each end, was the best system : the harbours to be formed by piers and dredging out to deep water.

"The Commission met at Paris in June 1856, when the views of the English engineers were, after full discussion, rejected, and the report to the Viceroy recommended the system which has since been carried out. The Commission estimated the work to cost £8,000,000.

"Two years from the date of this report were spent in conferences and preliminary steps before M. Lesseps obtained the necessary funds for carrying out the works. About half the capital was subscribed on the Continent, by far the larger portion being taken in France, and the other half was found by the Viceroy. Further time was necessarily lost in preparation, and it was not till near the close of 1860 that the work was actually commenced. . . .

"The original concession granted extraordinary privileges to the Company. It included or contemplated the formation of a 'sweet water' canal for the use of the workmen engaged, and the Company were to become proprietors of all the land which could be irrigated by means of this canal. One of the conditions of the concession also was that the Viceroy should procure forced labour for

the execution of the work; and soon after the commencement of operations, and for some time, the number of
workmen so engaged amounted to from 25,000 to 30,000.
The work thus commenced steadily proceeded until 1862,
when the late Viceroy during his visit to this country at
the time of the International Exhibition, requested Mr.
Hawkshaw to visit the canal and report on the condition of the works and the practicability of its being
successfully completed and maintained. His Highness's
instructions were that Mr. Hawkshaw should make an examination of the works quite independently of the French
company and their engineers, and report, from his own
personal examination and consideration, the result at
which he arrived. If his report were favourable the work
would be proceeded with, if unfavourable it would at once
be stopped.

  " Mr. Hawkshaw proceeded to Egypt upon this important commission in November of the same year, and in
February 1863 he wrote a well-considered report, which
may be said to have in a great measure contributed to
the rapid and successful completion of the work. Mr.
Hawkshaw described the works of the canal which had
been already executed and those which remained at that
time unfinished. He examined and discussed the dimensions of the various parts then in progress, recommending
various alterations, and having carefully gone into all the
details of construction, he proceeded to investigate the
question of maintenance, with reference to which it had
been urged by opponents

  " ' 1°. That the canal will become a stagnant ditch.

" ' 2°. That the canal will silt up, or that the moving sands of the Desert will fill it up.

" ' 3°. That the Bitter Lakes through which the canal is to pass will be filled up with salt.

" ' 4°. That the navigation of the Red Sea is dangerous and difficult.

" ' 5°. That shipping will not approach Port Saïd, because of the difficulties that will be met with, and the danger of that port on a lee shore.

" ' 6°. That it will be difficult, if not impracticable, to keep open the Mediterranean entrance to the canal.'

" Having analysed each of these objections, and fully weighed the arguments on which they were based, he came to the following conclusions as to the practicability of construction and maintenance :—

" ' 1st. As regards the engineering construction, there are no works on the canal presenting on their face any unusual difficulty of execution, and there are no contingencies that I can conceive likely to arise that would introduce difficulties insurmountable by engineering skill.

" ' 2dly. As regards the maintenance of the canal, I am of opinion that no obstacles would be met with that would prevent the work, when completed, being maintained with ease and efficiency, and without the necessity of incurring any extraordinary or unusual yearly expenditure.'

" The whole of Mr. Hawkshaw's report is well worthy of perusal, and I must congratulate him on the sound conclusions at which he arrived, and on the foresight by which he was enabled to point out difficulties and contin-

gencies which have since arisen. Could he at that time
have seen the full realization of the work he would
scarcely have altered the report he wrote.

"Saïd Pasha died between the period of Mr. Hawkshaw's examination of the country and the date of his
report. He was succeeded by his brother, Ismaïl, the
present Viceroy or Khedive, who, alarmed at the largeness
and uncertainty of the grants to the Canal Company, of
the proprietorship of land which could be irrigated by the
sweet-water canal, and anxious to retire from the obliga-
tion of finding forced labour for the construction of the
works, refused to ratify or agree to the concessions
granted by his brother. The whole question was referred
to the arbitration of the Emperor of the French, who
kindly undertook the task, and awarded the sum of
£3,800,000 to be paid by the Viceroy to the Canal Company as indemnification for the loss they would sustain
by the withdrawal of forced or native labour, for the
retrocession of large grants of land, and for the abandon-
ment of other privileges attached to the original act of
concession. This money was applied to the prosecution
of the works.

"The withdrawal of native labour involved very important changes in the mode of conducting the works, and
occasioned at the time considerable delay. Mechanical
appliances for the removal of the material, and European
skilled labour, had to be substituted; these had to be
recruited from different parts of Europe, and great difficulty was experienced in procuring them. The accessory
canals had to be widened for the conveyance of larger

dredging-machines, and additional dwellings had to be provided for the accommodation of European labourers. All
these difficulties were overcome, and the work proceeded."

After the works had been nearly completed, the Lords
of the Admiralty instructed Admiral Richards, the hydrographer, and Lieutenant-Colonel Clarke, R.E., to visit
Egypt, and report as to the condition of the canal. These
officers accordingly made a most minute survey of the
canal and its terminal harbours, and issued a most interesting report,[1] from the information contained in which
the plan of the canal, Plate III., has been mainly constructed. On referring to this plan it will be seen that
the canal extends from Port Saïd on the Mediterranean
to Suez on the Red Sea, and that, as shown by the
section, it traverses a comparatively flat country. This
route has been selected so as to take advantage of certain
valleys or depressions which are called lakes, but were in
fact, previous to the construction of the canal, low-lying
tracts of country, at some places below the level of the
Mediterranean and Red Seas. These valleys were found
to be coated with a deep deposit of salt, and are described
as having had all the appearance of being covered with
snow, bearing evidence of their having been at one period
overflowed by the sea. As will be seen from the plan,
Lake Menzaleh is next to the Mediterranean, Lake Timsah
about half-way across the isthmus, and the Bitter Lakes
next to the Red Sea. Lake Timsah, which is about 5
miles long, and the Bitter Lakes about 23, were quite dry

[1] Report on the Maritime Canal connecting the Mediterranean at Port Saïd
with the Red Sea at Suez, February 1870.

before the cutting of the canal, and the water which has converted them into large inland lakes was supplied from the Red Sea and Mediterranean. The water began to flow from the Mediterranean in February 1869, and from the Red Sea in July, and by the beginning of October of the same year these vast tracts of country, which had formerly been parched and arid valleys, were converted into great lakes navigated by vessels of the largest class. It will be seen from the section that the surface of the ground is generally very low, the chief cuttings being at Sérapéum and El Guisr, where the sandy dunes attain an elevation of about 50 to 60 feet. The channel through the lakes was excavated partly by hand labour and partly by dredging, and for a considerable portion the level of the valleys was such as to afford sufficient depth without resorting to excavation. The material excavated appears to have been almost entirely alluvial, and easily removed; the only rock was met with at El Guisr, where soft gypsum occurred, removable to a considerable extent by dredging, so that the canal works may be regarded as having presented no physical difficulty.

The following details as to the dimensions of the work are chiefly supplied from the Admiralty report, already referred to.

The whole length of the navigation is 88 geographical miles; of this distance 66 miles were actual canal, formed by cuttings, 14 miles were made by dredging through the lakes, and 8 miles required no works, the natural depth being equal to that of the canal. Throughout its whole length the canal was intended to have a navigable depth of

26 feet for a width of 72 feet at the bottom, and to have a width at the top varying according to the character of the cuttings. At these places where the cuttings are deep, the slopes were intended to be 2 to 1, with a surface width at the water-line of about 197 feet, as shown in fig. 10, which is a cross section at El Guisr; in the less elevated portions of the land, where the stuff is

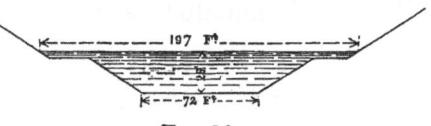

FIG. 10.

softer, the slopes were to be increased, giving a surface width of 325 feet. Of course it will be understood that in the lakes the canal consists of a navigable channel of sufficient depth and breadth to admit the traffic, the surface of the water extending on either side to the edge of the lake. Fig. 11 shows a cross section at Lake Manzaleh. At the date of the Admiralty inspection, these dimensions had not in all respects been fully attained, the depth at some places varying

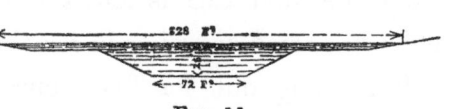

FIG. 11.

from 20 to 22 feet, but the Admiralty officers reported that the deepening of the shallows is in progress, and that they are likely soon to be removed. The curves, they also report, are sharp, requiring great care and attention in piloting vessels. The deep channel through the lakes is marked by iron beacons on either side, 250 feet apart, and the Admiralty reporters state that "in practice it is found more difficult to keep in the centre while passing through these beacons, than it is when between the embankments." At every five or six miles there is a passing-place, to enable large vessels to moor for the night, or to bring-up, in order to allow others to pass. At each of them a telegraph station is to be established, with an officer who is to regulate the movements of passing vessels according to directions which will be communicated from Port Saïd, Ismailia or Suez.

Perhaps the most interesting question to the engineer is that which relates to the action of the tide between the two seas, and in so far as observations have been made, they are given in the following quotation from the Admiralty report :—" The tidal observations which we were able to make were necessarily somewhat imperfect from want of time, but they were made at that period of the moon's age when their effect would be greatest; the results show that in the southern portion of the canal, between Suez and Great Bitter Lake, the tidal influence from the Red Sea is felt, there being a regular flow and ebb; the flood running in for about seven hours, and the ebb running out for five hours; at the Suez entrance, the rise at springs, unless effected by strong winds, is between 5 and 6 feet; about half way from Suez to the Small Bitter Lake, a distance of 6 miles, it is under 2 feet; at the south end of the Small Bitter Lake, a few inches only; while at the south end of the Great Lake there is scarcely any perceptible tidal influence. We were informed by the authorities at Ismailia, that since the Great Lake has been filled, the level of Lake Timsah, which was filled from the Mediterranean in April 1867, has risen 12 centimetres, or about 4 inches; and that its waters are continually running at a slow rate into the

Mediterranean ; certainly this statement agreed with what we ourselves remarked, for we always found a current running northward from Lake Timsah at the rate of from half a mile to a mile an hour. Limited, however, as these tidal observations were, they were taken with great care, and appear sufficient to show that, except at the Suez end, the tides will not materially affect the passage of vessels ; at that end, therefore, large vessels must regulate their time of passing; indeed, the greatest difficulty which will be experienced will be not from the tides but from the prevailing north-east wind in the canal, which will make close steerage difficult in going from north to south."

It would be highly interesting and valuable to have observations made simultaneously at various points, to ascertain the action of the tide. All that is at present known is contained in the Admiralty report, and appears to be, as already stated, that the tidal column of 5 feet range in the Red Sea is reduced to 2 feet at the distance of 6 miles, and is practically annihilated by the wide expanse of the Bitter Lakes.

In executing this strange work of the desert, and converting dry sands into navigable lakes, it is stated that there have been about *eighty* millions of cubic yards of material excavated, and at one time nearly 30,000 labourers were employed on the works. For their use a supply of fresh water was conveyed from the Nile at Cairo, and distributed along the whole length of canal. This work was one of no small magnitude. The fresh-water conduit is an open channel from Cairo to Ismailia, and thence to Suez the water is conveyed in pipes. The surplus fresh water is applied to the irrigation of the adjacent country. The cost of the whole undertaking, including the harbours, is stated to have been £16,000,000, and it is expected that probably £300,000 may still be required for its full completion.

The terminal harbours are important adjuncts of this great work. That on the Mediterranean is Port Saïd, which is shown on Plate III. It is formed by two break-waters constructed of concrete blocks; the western one 6940 feet in length, and the eastern 6020 feet, enclosing an area of about 450 acres, with an average depth of only 13 or 14 feet, excepting in the channel leading to the canal, where the depth is 25 to 28 feet.

The entrance to the canal at Suez is also protected by a breakwater, and, in connexion with the harbour at this place, there are two large basins and a dry dock.

As regards the capabilities of the canal for navigation, the Admiralty reporters state that it is a convenient highway for all steam-ships or vessels towed, ranging between 250 and 300 feet in length, with 35 feet beam, and a draught of 20 feet; for vessels of larger class the canal is not so well adapted, and special arrangements would require to be made and enforced for the transit of large vessels. Even vessels of 400 feet long, with 50 feet beam and 22 feet draught, could be taken through, by adopting special precautions. A delay of three days is calculated on for the passage across from Port Saïd to Suez.

Many fears have been expressed as to the feasibility

of maintaining this artificial passage at a remunerative expense to its constructors, on account of its being silted up by the drift-sand of the desert, as well as that brought in by the tidal currents—the wasting of the soft banks by the passage of vessels—and the difficulty of obviating the silting up of Port Saïd by sand carried through the open work of which the breakwater is formed, and deposited in the area of the harbour. This treatise, however, is not the place to discuss the probable success or failure of engineering works; all I profess to do is to explain the principles on which these works are designed, and give examples of such as have been successfully completed; and certainly, whatever may be the future fate of the Suez Canal, either as an engineering work or as a speculation, all praise must be accorded to M. Lesseps and his staff, for having, in the face of great difficulty, successfully executed one of the most remarkable feats of modern engineering.

CHAPTER III.

THE COMPARTMENTS OF RIVERS DEFINED.

Compartments of rivers—Their physical characteristics described—Example of Dornoch Firth—Boundaries of compartments not always distinct—Different compartments require distinct engineering works for their improvement.

Difference
between Canal
and River
navigation. FROM what has been said, it will be seen that a canal is a work by which water is diverted from its natural course, and made to occupy a channel prepared for its reception, extending through the country for the transport of boats and vessels. Canal navigation is thus entirely *artificial* in its character. In this respect it differs from river navigation, which may be described as the art of using, for the purposes of inland communication, rivers flowing in their natural courses, and of applying means to render them subservient to the purposes of navigation in cases where the depth is limited, or where rapid currents exist. Our consideration of rivers must therefore necessarily comprehend a general sketch of their physical characteristics, and the laws of their motion, as a necessary introduction to the practical part of the subject, which deals with the engineering works required for their improvement.

The compart-
ments described
as occurring
in rivers. As introductory, therefore, to the remarks which are to follow, it seems desirable to premise, as described by

me in 1845, in a communication to the Royal Society of Edinburgh,[1] that in all rivers affected by tidal influence, two physical boundaries, more or less apparent, according to circumstances to be afterwards noticed, are invariably found to exist, caused by the influx of the tidal wave through firths or bays, and the modification it receives in its passage up the gradually rising inclination or slope of a river's bed. These boundaries again produce three compartments.[2] The seaward, or lowest of these, I termed the " *sea proper;*" the next, or intermediate one, into which the sea ascends, and from which it again withdraws itself, I termed the " *tidal compartment of the river;*" and the highest, or that which is above the influence of the sea, the "*river proper.*" Their relative extent in different situations is influenced not only by the circumstances under which the great tidal wave of the ocean enters the river, but by the size of its stream, the configuration and the slope of its bed, and, in short, by every natural or artificial obstruction which is presented to the free flow of the tidal currents along its channel.

These three compartments possess very different phy- Their physical characteristics. sical characteristics. The presence of *unimpaired tidal phenomena* in the lowest, the *modified flow of the tide,* produced by the inclination of the river's bed in the intermediate, and the *absence of all tidal influence* in the highest compartment, may be shortly stated as the phenomena by which these spaces are to be recognised. The

[1] *Proceedings of the Royal Society of Edinburgh,* vol. ii. p. 26.
[2] Lord Cockburn, in addressing a jury in 1837, appears to have stated similar boundaries.

tides in the " sea proper" compartment of an estuary,
for example (although the place of observation be several
miles embayed from what in strictness could be called
the " sea" or " ocean"), will be found to resemble those
of the adjoining sea with which it communicates,—*First*,
in the identity of the levels of low water ; *second*, in the
shortness of the time which elapses between the cessation
of ebbing and the commencement of flowing, or, in other
words, the absence of any protracted period of low water,
during which the surface appears to remain stationary at
the same level ; *third*, in the symmetrical form traced by
the passage of the tidal wave ; and *fourth*, in the range of
tide, so far as that is not influenced by the formation of the
shores in narrow seas or channels. In ascending into the
intermediate compartment, however, the level of the low
water is no longer the same ; the range of tide, excepting
in peculiar cases, becomes less, and is gradually decreased
as the bed of the river rises, and at length a point is
reached where its influence is not perceptible. In this
intermediate section the phenomena of *ebbing* and *flowing*
are still found to take place, but the times of ebb and
flow do not remain constant, that of ebb gradually gaining the ascendency, and in some cases never entirely
ceasing, though the level of the river be rising. The
duration of low water is also gradually protracted as we
proceed upwards, until the influence of tide is unknown.
This forms the boundary-line of the upper compartment,
the characteristic of which is the total absence of ebbing
and flowing ; the river at all times pursuing its downward course in an uninterrupted stream, and at an. un-

PLATE IV.

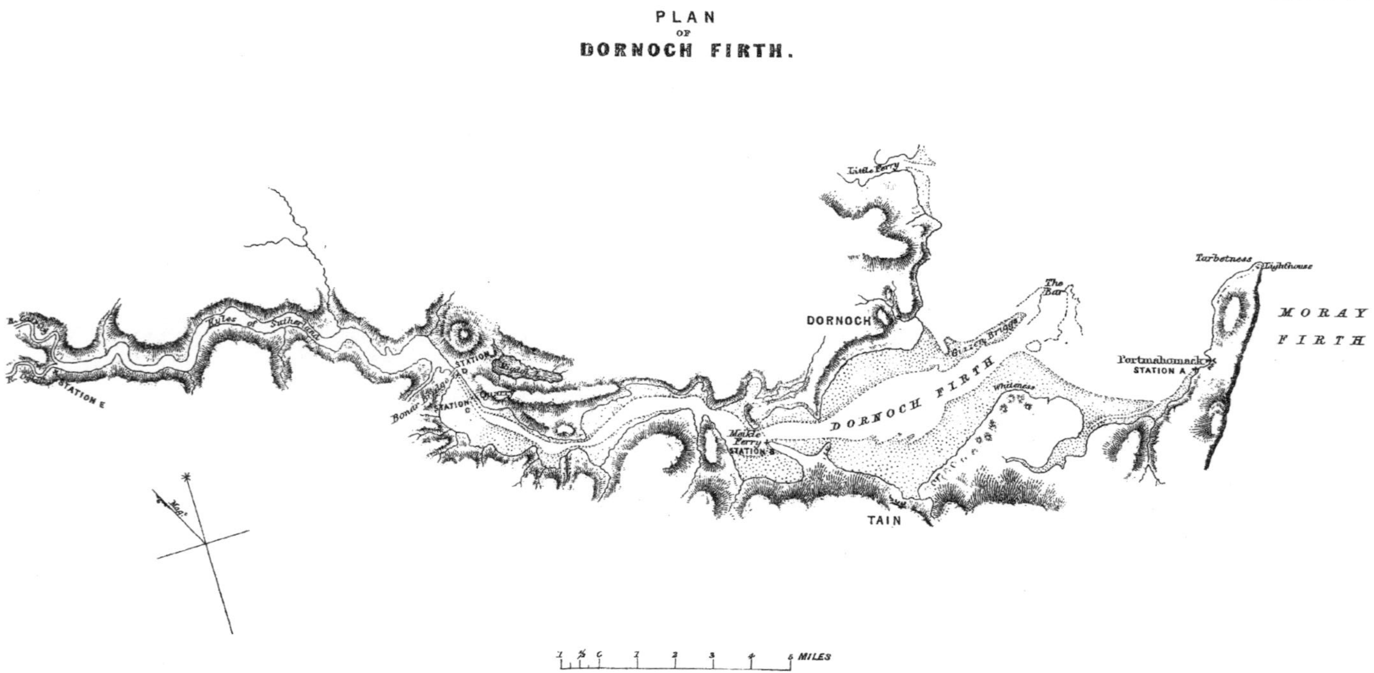

PLAN
OF
DORNOCH FIRTH.

MORAY FIRTH

Tarbetness Lighthouse

Portnahomack
STATION A

The Bar

DORNOCH FIRTH

DORNOCH

Ferry Town

DORNOCH

Little Ferry

STATION B

TAIN

STATION D

STATION E

N

1 ½ 5 ¼ 1 2 3 4 5 MILES

Published by A & C. Black, Edinburgh.

varying level, except in so far as may result from the changes due to land floods.

In the investigation of these different characteristics, the variable nature of the elements to be dealt with must be kept in view. The river, for example, is liable to be affected by floods, and the state of the tides by winds and other causes ; and therefore a great degree of precision in defining these spaces cannot in all cases be expected. But the termination of the low-water level at the separation of the seaward and intermediate spaces, as laid down by marine surveyors, simply from observation of the tidal phenomena, has in several situations been found to agree exactly with the position of that boundary as determined by engineers by means of accurate levelling, combined with careful tidal observations.

An example in actual practice will best illustrate what is meant, and for this purpose I shall refer to the investigation of the tidal phenomena of the Firth of Dornoch Tidal phenomena of Dornoch Firth. and Kyles of Sutherland in Cromartyshire, made by me in 1842. By referring to the small chart of the Dornoch Firth in Plate IV. the reader will be better able to follow the illustrations to be given. The harbour of Portmahomac, marked A on the chart, about 3 miles from Tarbetness Lighthouse, was selected as the place at which to observe the *ocean* or *sea wave*. The second station at which it was found convenient to institute observations was within the Firth at Meikleferry, marked B, about 3 miles above the town of Tain, and 11 miles distant from Portmahomac. The third station was at Bonar Quarry, marked C, situated on the north shore of the Firth, and

8 miles inland from Meikleferry; and the fourth station was at Bonar Bridge, marked D, one mile from the Bonar Quarry. Beyond Bonar Bridge the observations were also extended as far as the junction of the rivers Oykell and Cassley, marked E, a distance of $12\frac{1}{2}$ miles; so that the whole distance embraced in the investigation was $33\frac{1}{2}$ miles. Graduated tide-gauges were fixed at Portmahomac, Meikleferry, Bonar Quarry, and Bonar Bridge; and by means of two distinct sets of levelling observations, the heights of the zeros of these gauges, in relation to each other, were accurately determined, so that all the tidal observations made at them could be reduced to the same *datum* line. The result of the tide observations was, that *the low water of each tide is, practically speaking, on the same level at Portmahomac, Meikleferry, and Bonar Quarry.* I use the word *practically*, because the level of the sea is more or less affected by every breeze of wind, which necessarily must *pen up* and elevate some portions of its surface, and cause corresponding depression at other places, so that an unvarying low-water level will not be found to exist throughout a series of tides on any part even of the ocean itself, however limited the number of low waters embraced may be. Accordingly, deviations of a few inches from the true level occasionally occurred in the observations made at the Dornoch Firth; but these were not of greater extent than could reasonably be traced to the effect of wind, and were found to vary, not only in their amount, but also in their value, being at the same gauges sometimes *plus* and sometimes *minus* quantities, causing corresponding varia-

Low-water line practically level from Portmahomac to Bonar Quarry.

tions in the results deduced from the different series of tidal observations that were made. Some of these showed the low water within a fraction of an inch of being level ; while others gave a notable *elevation* at some of the stations ; and others, again, gave a *depression* below the level line at the very stations where previously there had been a rise.

To illustrate this more fully, I shall give a few examples : Thus, on the 23d of June (on which day the weather happened to be very calm), the level of low water at Meikleferry was three-quarters of an inch *above* that at Portmahomac ; and on the next day, the wind blowing fresh from the s.e., the level of low water at Meikleferry was $3\frac{3}{4}$ inches *above* that at Portmahomac. Again, a succeeding observation gave the level of Meikleferry three-quarters of an inch *below* Portmahomac. In the same way, and in similar small degrees, the level between the low water at Bonar Quarry tide-gauge and at Portmahomac was found to vary. The *average* of all the observations made gave the level of low water at Meikleferry 2·2 inches *above* that at Portmahomac, and the level of low water at Bonar Quarry 1·1 inch *below* the low water at Portmahomac. Whether these *average* differences of level be traceable to the effects of prevailing winds, which may be supposed to have exerted a greater influence on the water at the more exposed stations, or to any inaccuracy in the levels, must evidently, from the examples given of the extent and nature of the daily deviations, be a point which we cannot determine ; but the result of a lengthened train of observations, notwithstanding the *average* difference above stated, may fairly be held to be, that the low water of each tide is *practically* on the same level at Portmahomac, Meikleferry, and Bonar Quarry; and therefore that the *low-water* tidal phenomena, throughout the whole extent of the firth, correspond with those of the sea.

Low water rises from Quarry to Bonar Bridge. But when the results of the observations at Bonar Bridge come to be compared with those made at the station immediately seaward of it, a very marked difference presents itself; for, while the low-water line is found to be practically level from Portmahomac to Bonar Quarry, a distance of 20 miles, throughout a narrow firth, varying from 1¼ mile to 550 feet in breadth at low water, we find that between the Quarry and Bonar Bridge, a distance of only 1 mile, there is a rise in the low-water line of spring-tides of no less than 6 feet 6 inches. It was therefore concluded that, in the Dornoch Firth, the point at which the low-water level of spring-tides met the descending current of fresh water lay somewhere between the Quarry and Bonar Bridge. A different series of observations was made to ascertain the exact point at which this junction takes place, and the result of these observations was, that at low water of an ordinary spring-tide, rising 14 feet at Meikleferry, the low-water level of the sea meets or intersects the descending fresh-water stream from the Kyle of Sutherland, at a point 1700 yards below Bonar Bridge, or nearly opposite Kincardine Church, and within 60 yards of the Quarry station. Between this point and the Bridge, a distance of 1700 yards, there is a rise of 6 feet 6 inches, giving an average slope on the surface of the river of 1 in 784, or 6·7 feet per mile.

In addition to this uniformity in the level of low water, it was further found that the tidal phenomena of the firth corresponded to that of the adjoining sea, in the outline traced by the passage of the tidal wave, as deduced from observations made at the different stations on the rise and fall of the tide-level between the periods of low and high water.   During the period between each low water or high water the level of the surface was ever varying, there being no lengthened cessation of ebbing and flowing, the tide-wave being fully developed at the whole of the stations up to Bonar Quarry.   The range of tide was indeed increased in the inner part of the firth to the extent of 9 inches at Meikleferry, and 12 inches at Bonar Quarry ; that is, when the range of tide was 12 feet 8 inches at Portmahomac, it was 13 feet 5 inches at Meikleferry, and 13 feet 8 inches at Bonar Quarry.

But if we inquire into the tides at Bonar Bridge, we find that they do not correspond with those of the adjoining sea or of the firth ; for, taking the tide to which we have already alluded, which rose 13 feet 8 inches at Bonar Quarry, it was found on the same day to rise only 6 feet 10 inches at Bonar Bridge ; the difference between the two results being occasioned by the rise on the low-water line of the channel between these two places.   The tide on the particular day alluded to rose 6 feet 10 inches at Bonar Quarry before it affected the gauge at Bonar Bridge, when it began to rise at that place also, and afterwards continued to flow nearly uniformly at both places.

Rise of tide at Bonar Bridge.

Fig. 12 is a diagram illustrative of the form of the tide-wave at Meikleferry and Bonar Bridge : the hard line

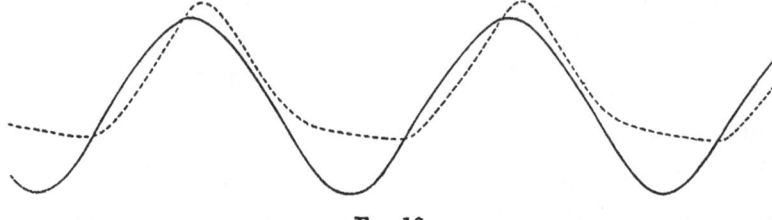

represents the curve formed by the passage of the tidal wave at Meikleferry, and the dotted line shows that at Bonar Bridge. In both cases the vertical space represents the rise of tide, and the horizontal space the elapsed time. From this diagram it will be seen, that while the tide at Meikleferry is symmetrical, and presents a constantly rising or falling outline, the tide at Bonar Bridge represents a long period, extending on some occasions by actual observation to several hours at low water, nearly unaffected by tidal influence, during which period the water stood almost at the same level. The tidal water admitted into the upper part of the estuary above Bonar Bridge took a considerable time to drain off through the narrow water-way at that place, and hence the water did not attain a permanent low-water level, even long after the tide had ceased to operate in affecting its surface. The observations made to ascertain how far the tidal influence extended up the Kyles of Sutherland were conducted with the same care, and proved that the highest point influenced by the tide was at the junction of the rivers Oykell and Cassley, 12½ miles above Bonar Bridge. In stating, however, that the low waters at Portmahomac,

Meikleferry, and Bonar Bridge are on the same level, the reader must not infer that the surface of the water throughout the firth presents at any period a *level plane*. Although the low waters are identical as regards level, the times of low water are not the same. On the contrary, the time of low water at Meikleferry was sometimes found to be 50 minutes later than at Portmahomac, and that at Bonar Quarry 50 minutes later than at Meikleferry, so that when the water had attained its lowest level at Bonar Quarry it had been rising for 1 hour and 40 minutes at Portmahomac ; there is, therefore, at no period a level plane extending throughout the firth, but what may be termed a constantly undulating surface. This will be better understood when we come to treat of tide observations, and to show their results as obtained on different rivers.

A further test of the " sea proper " will, it is believed, Mean sea-level. be found in the existence, at any place of observation within that compartment, of a central point in the vertical range of tide from which the high and low water levels of every tide are very nearly equidistant. The existence of such a point was, it is believed, first determined by Mr. James Jardine, at the Tay, in 1810,[1] and has been observed in the firths of Forth and Dornoch, at the Skerryvore Rocks on the west of Scotland, at the Isle of Man, and in the Mersey. These different series of observations, made at points so far distant from each other, seem to prove the universality of the phenomenon, at least on the shores of this country. But in ascending

[1] Report by James Jardine, C.E.

into the tidal compartment, the rise on the low-water level, which has already been described, destroys at once the symmetry of the tide-wave, as shown in fig. 12, and the existence of any such central point equidistant from the high and low water level of each tide.

The case I have adduced serves to illustrate the definition I have given of the compartments of rivers. From Portmahomac to Kincardine, near Bonar Quarry, we have all the evidences of what I have termed the "sea proper;" the line traced through the low-water mark at different parts of the firth is practically level; the curve formed by the rise and fall of the tide is symmetrical; there is no lengthened cessation of ebbing and flowing at the period of low water, and the range of tide is unmodified save by the additional rise due to the narrow firth through which the tide-wave passes. From Kincardine to the junction of the Oykell and Cassley we have proofs no less evident of the modified flow of the tide peculiar to the "tidal compartment." Even at Bonar Bridge, 1 mile above the Quarry, the low-water level is 6 feet 6 inches higher than at the station below. At low water the tide remains within a few inches of the same level for several hours, and its maximum range is reduced to about one half of what it is further seaward, while at the junction of the Oykell and Cassley it disappears altogether. Above this point no tide is known to affect the flow of the stream, which, being free from all tidal influence, may be termed the "river proper."

These boundaries not equally distinct in all cases. I must here warn the reader not to suppose that the boundaries we have traced as existing in the Dornoch

Firth, and other places which I have investigated, may be determined with the same precision under all circumstances and in every case. The observations to which I have alluded are supposed to be made at periods when the river is free from floods and the sea unaffected by heavy gales; moreover, the configuration of the bottom and shores of a river and estuary may, in certain cases, render the accurate determination of the boundaries very difficult. All that I state is, that these compartments do in some measure, more or less defined, exist in all rivers debouching through firths and estuaries on a coast where the sea has a notable range of tide. These are the *general features* of the rivers in this country, but in places where the range of tide is barely perceptible, such as the Mediterranean, or where the river joins the ocean by a short and steep descent, as will be afterwards noticed, the boundaries I have defined cannot be easily traced.

The three compartments which have been defined as existing in rivers and estuaries naturally lead to a convenient division of our subject in treating of River Engineering; for it so happens that the physical characteristics described as peculiar to each of the compartments are not less distinct than the engineering works required for their improvement. Thus, for example, on the "river proper" section the works may be said, in general terms, to consist chiefly of weirs built across the stream, by which the water is dammed up and forms stretches of canal in the river's bed, with cuts and locks connecting the different reaches. The "tidal compartment" embraces a more

Different compartments require distinct engineering works for their improvement.

varied range of work, including the straightening, widening, or deepening of the courses and beds of rivers—the formation of new cuts, the erection of walls for the guidance of tidal currents, and, in some cases, the shutting up of subsidiary channels,—while the "seaward compartment" embraces such works as have for their object the improvement or removal of bars and shoals. All of these works will be treated in succeeding chapters.

It is necessary, however, to explain that no engineer is able to consider and advise as to the improvement of any portion of a river or estuary unless he is furnished with such data for his guidance as can be acquired only by accurate surveys and observation of its physical characteristics. Moreover, as it is impossible for the engineer, without possessing these data, to design improvements, so, I apprehend, it will greatly assist the student of Engineering if, before describing river-works, and showing their application in practice, I should give a brief sketch of some of the most important hydrometric investigations connected with river engineering. These include, among other things, accurate tidal observations and soundings,—the determination of a river's slope, velocity, and discharge,—the nature of its bed and banks, and other cognate inquiries, and I shall confine my notice to such of those topics as are likely to be most useful in illustrating or rendering more intelligible the subjects treated in the succeeding chapters, and must refer the reader for full particulars as to the character and extent of such information, and the *means of accurately obtaining* it, to works on River and Marine Surveying.[1] Neither do I propose to include in the present treatise any account of the interesting and gradual progress made by philosophers and engineers of the early Italian and French schools in theoretical and experimental investigations of the laws which regulate the flow of water in natural and artificial channels, and form the groundwork of all our practice in hydraulic engineering. These lengthened and laborious experimental researches will be found most fully discussed—historically, theoretically, and practically—in the article by Dr. Robison, on the "Theory of Rivers," in the *Encyclopædia Britannica,* and also in the reports made by Mr. George Rennie to the British Association "On the State of our Knowledge of Hydraulics as a branch of Engineering,"[2] to which I refer the reader.

[1] *Application of Marine Surveying and Hydrometry to the Practice of Civil Engineering.* By David Stevenson, Civil Engineer. Edinburgh, 1842, A. & C. Black.

[2] *Reports of British Association for Advancement of Science,* 1833 and 1834.

CHAPTER IV.

HYDROMETRIC OBSERVATIONS.

Tides of rivers—Variations in the tidal lines—Professor Robison's remarks on the anomalies of river tides—Nature of the inquiry into river tides—Tide-gauges—Selection of stations for tide observations—Agents which produce disturbance in the parallelism of the tidal lines—Manner in which these variations affect the soundings—Datum line for soundings—Use of tide-gauges in reducing soundings to the datum—Formula for their reduction—Formula only true on the supposition of the lines being parallel to high water—Results affected by erroneous supposition—Most effectual means of avoiding inaccuracy; but this not always practicable—General rules for taking soundings—High and low water soundings—Formulæ for ascertaining the rise of tide and height of sandbanks—Cross sections—Mr. Henry Mitchell's rule for determining elevations along a tidal river without levelling; not applicable to rivers in this country.

Tidal observations.

THE hydrometric observations to be first noticed are those of the tides, and it is essential that they be scrupulously correct, as they determine the accuracy of all soundings of the depth of water, and of all sections of the bed of the river, which are generally constructed from the data afforded by the sounding-line. The information afforded by the tidal and other hydrometric observations is also indispensable, as will afterwards be seen, in enabling the engineer to form an opinion on many important questions that are being constantly brought before him, and I therefore offer no apology for discussing this part of the subject at some length.

The tides of rivers are influenced partly by the circumstances under which the great tidal waves of the ocean enter their mouths or estuaries, and partly by the size of the streams and the configuration of the beds and banks of the rivers themselves, all of which have a share in modifying the free flow of the tidal currents along their channels. As no rivers are to be met with whose communication with the sea, and the course and strength of whose streams are in all respects similar, corresponding dissimilarities naturally occur in the circumstances attending the rise and fall of river tides. If it were correct to assume that the high-water mark of each tide, at any given number of points in a river's course, stood invariably on the same level,—that the times of high water at these points were the same,—and that the progressive rise and fall of the tides were uniform and equal at every point,—or, in other words, that the sectional lines formed by the surface of the water at all periods of flood and ebb, which I term the "tidal lines of the river," were parallel to the line of high water,—the work of the engineer in acquiring data would be greatly simplified. But if he were to make such an assumption the groundwork on which to found his opinions and frame his designs, his conclusions would almost invariably be formed on erroneous data, as his soundings and sections would in most cases be inaccurate.

The following remarks by Professor Robison,[1] illustrative of some of the anomalies in river tides, are interesting in connexion with this subject.

Anomalies of river tides.

Regarding the rise or inclination which in certain circumstances occurs in the high-water line, from the

[1] Robison's *Mechanical Philosophy* (Brewster's Edition), vol. iii. p. 353.

entrance of a river upwards, Dr. Robison says :—" When a wave of a certain magnitude enters a channel, it has a certain quantity of motion, measured by the quantity of water and its velocity. If the channel, keeping the same depth, contract its width, the water, keeping for a while its momentum, must increase its velocity or its depth, or both, and thus it may happen that, although the greatest elevation produced by the joint action of the sun and moon in the open sea does not exceed 8 or 9 feet, the tide in some singular situations may mount considerably higher. It seems to be owing to this that the high water of the Atlantic Ocean, which at St. Helena does not exceed 4 or 5 feet, setting in obliquely on the coast of North America, ranges along that coast in a channel gradually narrowing till it is stopped in the Bay of Fundy as a hook, and there it heaps up to an astonishing degree." Again, as to the variation in the times of high water at different points, and the non-parallelism of the tidal lines, he says :—" Suppose a great navigable river, running nearly in a meridional direction, and falling into the sea in a southern coast. The high water of the ocean reaches the mouth of the river (we may suppose) when the sun and moon are together in the meridian. It is therefore a spring-tide high water at the mouth of the river at noon. This checks the stream at the mouth of the river, and causes it to deepen. This again checks the current farther up the river, and it deepens there also, because there is always the same quantity of land water pouring into it. The stream is not perhaps stopped, but only retarded. But this cannot happen without its grow-

ing deeper.   This is propagated farther and farther up
the stream, and it is perceived at a great distance up the
river.   But this requires a considerable time.   We may
suppose it just a lunar day before it arrives at a certain
wharf up the river.   The moon at the end of the day is
again on the meridian, as it was when it was spring-tide
at the mouth of the river, the day before.   But in this
interval there has been another high water at the mouth
of the river, at the preceding midnight, and there has just
been a third high water about 15 minutes before the
moon came to the meridian, and 35 minutes after the sun
has passed it.   There must have been two low waters in
the interval at the mouth of the river.   Now, in the same
way that the tide of yesterday noon is propagated up the
stream, the tide of midnight has also proceeded upwards,
and thus there are three co-existent high waters in the
river.   One of them is a spring-tide, and it is far up at
the wharf above mentioned.   The second, or the midnight
tide, must be half-way up the river, and the third is at
the mouth of the river.   And there must be two low
waters intervening.   The low water, that is, a state of
the river below its natural level, is produced by the passing low water of the ocean, in the same way that the
high water was.   For when the ocean falls below its
natural level at the mouth of the river, it occasions a
greater declivity of the issuing stream of the river.   This
must augment its velocity ; this abstracts more water
from the stream above ; and that part also sinks below
its natural level, and gives a greater declivity to the
waters behind it.   And thus the stream is accelerated,

and the depth is lessened in succession, in the same way
as the opposite effects were produced. We have a low
water at different wharfs in succession just as we had
the high waters."

"This state of things, which must be familiarly known
to all who have paid any attention to these matters,
being seen in almost every river that opens into a tide-
way, gives us the most distinct notion of the mechanism
of the tides. It is a great mistake to imagine that we
cannot have high water at London Bridge (for example),
unless the water be raised to that level all the way from
the mouth of the Thames. In many places that are far
from the sea, the stream at the moment of high water is
down the river, and sometimes it is considerable. At
Quebec it runs downwards at least 3 miles per hour.
Therefore the water is not heaped up to a level, for there
is no stream without a declivity."

In the river Amazon, the tide is said to ascend against
the stream, in the manner described, for several days,
and to penetrate to the distance of 200 leagues from
its mouth, seven or eight tides, with intermediate low
waters, following each other in succession;[1] and in the
Thames we find a similar tidal succession, but not to so
great an extent, and arising, according to Whewell,
"from the peculiar circumstance of the river's having a
tide compounded of two tides arriving by different roads,
after journeys of different lengths," in allusion to the two
branches into which the tidal wave is divided on reaching
the British shores, one of which flows up the English

---

[1] *Encylopædia Britannica*, art. "River."

Channel, while the other proceeds along the west and northern coast of the country, and flowing down the east coast, again joins the other branch.

Such variations on the high-water lines as those described by Robison would no doubt be found to exist to some extent in every situation, if the rise of tide and the capacity of the river or estuary were sufficiently great to admit of their full development, and if the observations made were of sufficient extent to include them within their range. But from the smallness of British rivers, which flow from a comparatively narrow and contracted country, the ordinary surveys made for engineering purposes in Britain very rarely embrace so great a field of observations as to include the range of more than one tide; nevertheless, even in this country, such irregularities are found to exist on the tidal lines as to require careful investigation to insure accuracy, especially in situations where the rise of tide is great.

Before entering fully on the explanation of the different steps to be taken in making a correct series of tidal observations, by which alone the anomalies I have alluded to can be discovered, some preliminary remarks, in explanation of the exact nature of the inquiry to be instituted, appear necessary to the proper understanding of what is to follow.

If the tidal lines of a river were level and parallel, a Nature of the series of observations on the progressive rise and fall of inquiry into river tides. the tides made at a single graduated gauge placed in any part of its course, at which the whole of the tidal rise and fall is developed, would be sufficient for correcting all soundings and reducing them to one level datum line. For the person making the soundings on any part of the river would only have to note the time of observation, and by comparing it with the tide-observations made at the same time, as entered in the tide-book kept at the gauge, would discover the exact state of the tide at the time the sounding was made. If, on the other hand, the lines had a certain inclination, but were nevertheless parallel to each other, the single series of observations alluded to would still be sufficient for obtaining the correct depths at high water, and consequently an accurate profile of the bed of the river,

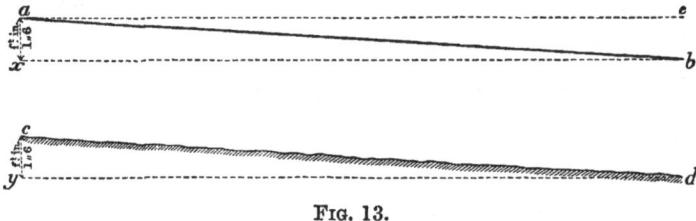

Fig. 13.

exhibiting all its inequalities; but it is evident that the inclination of the tidal lines, and, what is of more importance, the true position of the bed of the river in reference to the datum line of the section, could not be ascertained by this means. Thus let the lines $a\ b$ and $c\ d$ represent the high-water line and the bed of a river respectively, and let there be a rise of 1 foot 6 inches in both of them in the distance represented in the cut. If one tide-gauge only were used, suppose at the lower extremity of the river, the section, when protracted, would assume the form represented by the dotted lines $x\ b$ and $y\ d$, in which the high-water line and bottom of the river are shown as being level, whereas their correct positions in reference to the level line *a e*, which we may suppose to be the datum line of the section, are those represented by the lines *a b* and *c d*, on each of which there is a rise of 1 foot 6 inches.

The inclination of the bed forms an important element Tide-gauges. in all questions relative to the navigation of rivers, and proper means must be adopted for determining this before any design of improvement can be formed, and for this purpose it is obvious that at least two tide-gauges must be used, one at either extremity of the river; and further, that their relative levels must be accurately ascertained. Now, if the high-water line in the case referred to in fig. 13 should stand at 10 feet on the lower gauge, it will, *if their zeros are at the same level*, stand at 11 feet 6 inches on the upper one at the same moment, thus indicating the difference of level. In this way not only are the data for ascertaining the correct depths at high water afforded, but a proper section of the river can be made, its tidal lines and bed being represented in their true positions in reference to the datum line.

From what has already been said, however, regarding the anomalies of the tides, it will readily be seen that it would be improper to assume that the tidal lines are *parallel* during the whole period of flood and ebb; and therefore it is necessary to provide for this by adopting intermediate stations for tide observations, and by taking the soundings of the river at particular periods when the deviation from parallelism in the tidal lines is at its *minimum*, as will be more particularly noticed hereafter.

Selection of
stations for
tidal observa-
tions.
In determining the number and selecting the sites of
the stations at which tide observations are to be made,
the engineer ought to be regulated by the amount of
tidal rise and the configuration of the banks of the river.
Where the rise of tide is small, and the tidal currents are
very languid, few places of observation may suffice, but
where there is a great rise of tide, accompanied by rapid
currents, the parallelism of the tidal lines, on which the
correctness of the soundings depends, is more apt to be
disturbed, requiring a greater number of points of observation. It may be stated, as a general rule, that the more
numerous the tide stations are the nearer will the results
approximate to the exact line of the tidal wave at any
particular moment of flood or ebb, and the less chance
will there be of error in reducing the depths of the
soundings.

Agents which
produce dis-
turbance in the
parallelism of
tidal lines.
Whether an extended or limited series of observations
is to be adopted, it is necessary, while selecting the sites
for the stations, to have due regard to the agents most
likely to produce disturbance in the parallelism of the
tidal lines, such as abrupt turns or bends and sudden
enlargements in the transverse sectional areas of rivers.

These variations on the tidal lines, and the manner
in which they affect soundings and sections, will be best
explained by reference to a few examples.

Examples of
variations on
the tidal lines.
The results which I shall state, in the first place,
were obtained from observations made on the river Dee
in North Wales.

Three series of simultaneous tide observations were
made in that river: one at Chester, another at Connah's

PLATE V.

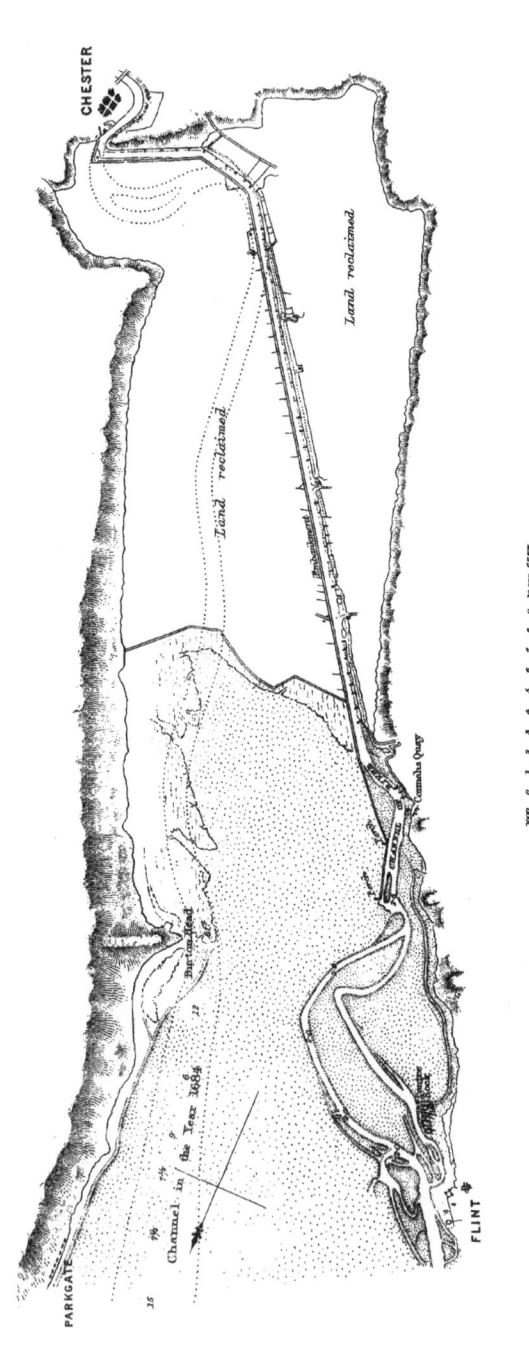

PLAN
OF PART OF THE
RIVER DEE.
1859.

CHESTER

Land reclaimed.

Land reclaimed.

Saltney Quay

QUAY

FLINT

PARKGATE

Channel in the Year 1684

Furtscotted

Published by A. & C. Black, Edinburgh.

Quay, and a third at Flint, and the following results were obtained.

The distance from Chester to Connah's Quay is 7⅓ miles, and that from Connah's Quay to Flint 3⅔ miles; the whole distance from Chester to Flint being 11 miles. The part of the river which extends from Flint to Connah's Quay may be said to be an open estuary; and the upper part, extending from Connah's Quay to Chester, is an artificial tidal canal, having an unobstructed waterway of about 500 feet in breadth at high, and 250 feet at low water, as will be seen from the chart of the river, Plate V.

The high-water line was found, by an average of twenty-four observations, to rise 2 inches from Flint to Connah's Quay; and from Connah's Quay to Chester the rise was found to vary from 4 inches at neap to 14 inches at spring tides, giving, as the result of twenty-four observations, an average rise of 6 inches. The whole average rise on the high-water line from Flint to Chester is therefore 8 inches.

The difference between the times of high water at the different stations was found to vary very much, and appeared to be more affected by the state of the winds than by the circumstance of the tides being neap or spring; but the average of the observations gave the time of high water at Flint twenty minutes earlier than at Connah's Quay, and that of high water at Connah's Quay thirty minutes earlier than at Chester; the whole average difference in time between high water at Flint and at Chester being fifty minutes.

The average level of the low-water line at Connah's Quay is 2 feet 6 inches below that at Chester, giving on the distance of 7⅓ miles an average fall of 4·09 inches per mile, and the level of the low water at Flint is 7 feet 6 inches below that at Connah's Quay, giving on a distance of 3⅔ miles an average fall of 24·54 inches per mile. The total fall from Chester to Flint is 10 feet, being an average fall on the distance of 11 miles of 10·9 inches per mile.

When the rise of tide, as indicated by the Liverpool tide-table, is 18 feet on the dock sill at Liverpool, the rise in the Dee is 20 feet 10 inches at Flint, 13 feet 8 inches at Connah's Quay, and 11 feet 5 inches at Chester.

Plates VI. and VII. represent approximately the forms assumed by the tidal lines of the river. Plate VI. represents the flood lines of a tide rising 19 feet 8 inches at Flint. In this, as well as in the other diagrams illustrative of the rise or fall of the tides, the perpendicular lines show the relative positions of the stations, and are graduated in the same way as the tide-gauges. On the horizontal line at the top of the diagrams, the relative distances between the stations are marked in miles, and at the right side of the Plates, the time corresponding to the level of the tide is expressed in hours and minutes. The hard diverging lines are drawn through the points at which the tide stood at the different stations, as ascertained by observation, and represent the tidal lines of the river. They are drawn straight, but in reality will present a curved form. Those which are *dotted* show the probable direction of those lines, when their forms could

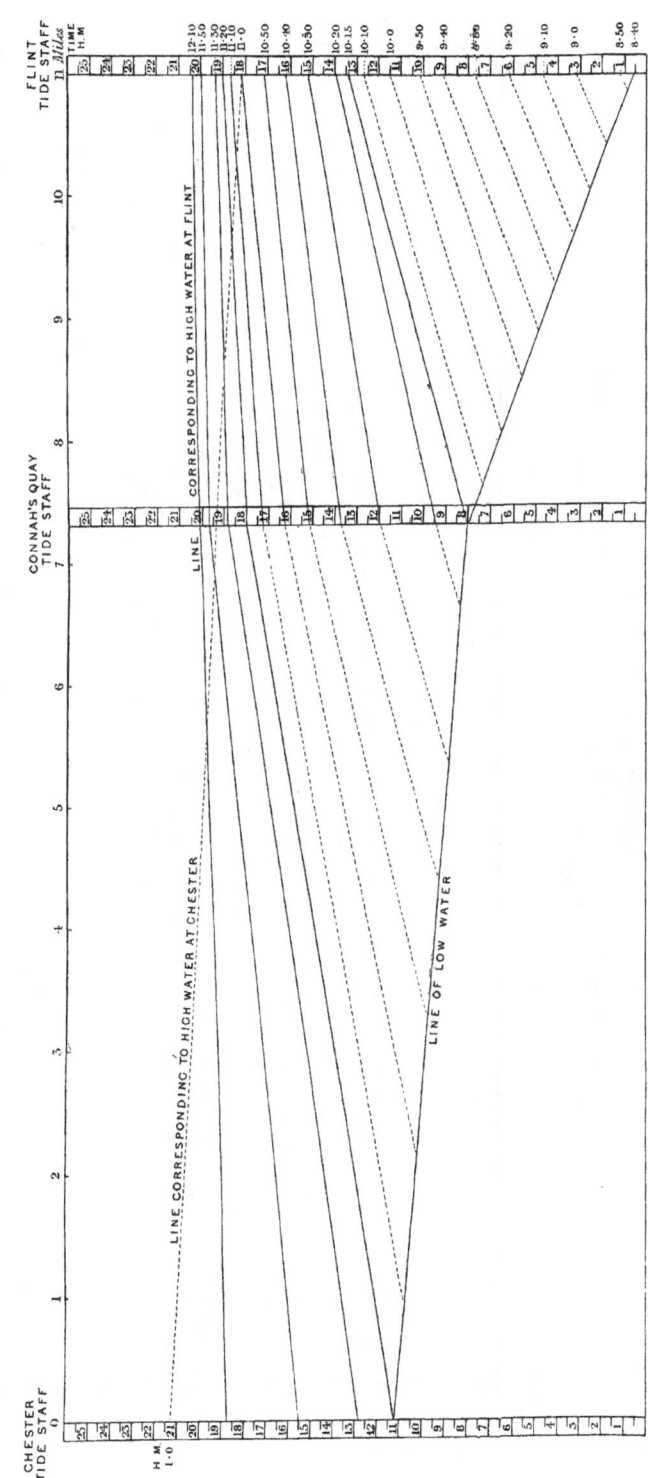

PLATE VI

RIVER DEE.
FLOOD OF A SPRING TIDE.

Diagram shewing the approximate forms assumed by the Tidal lines of the River Dee.
During the Flood of a Spring Tide rising 19 Feet 8 Inches at Flint.

James Andrews. Delt.

PLATE VII.

RIVER DEE.
EBB OF A SPRING TIDE.

Diagram shewing the approximate forms assumed by the Tidal lines of the River Dee,
During the Ebb of a Spring Tide rising 19 Feet 8 Inches at Flint.

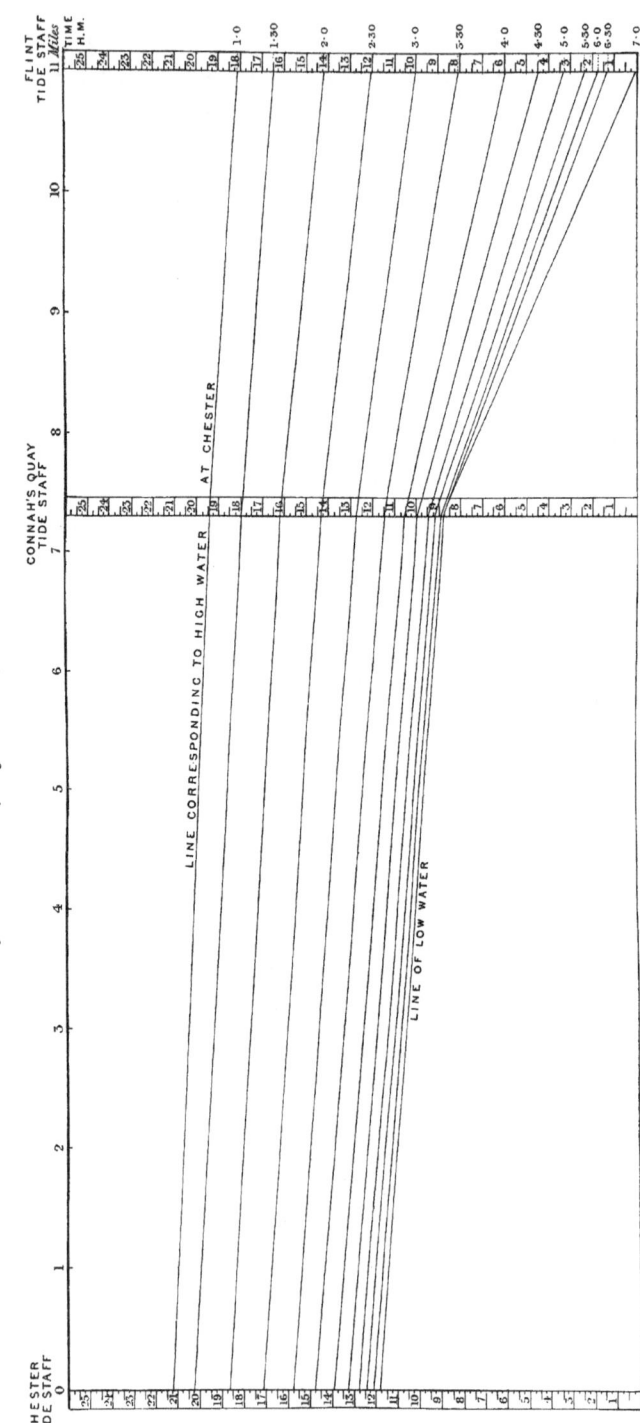

James Andrews, Delt.

not, for want of additional tidal stations, be more accurately determined.

The tide, as will appear from an inspection of Plate VI., began to rise at Flint at 8 hours 40 minutes; at 10 hours 15 minutes it had risen 12 feet 8 inches, and at that time had just appeared at Connah's Quay, the surface of the water at Flint being 5 feet 4 inches above that at Connah's Quay. At 11 hours 20 minutes the tide had risen 18 feet 4 inches at Flint, and was 1 foot above the level of the water at Connah's Quay, and 7 feet 10 inches above that at Chester, at which place the tide had just begun to appear. Thus, while at low water there is a fall of 11 feet from Chester to Flint, there was at the time above mentioned a fall of no less than 7 feet 10 inches on the surface of the water from Flint to Chester At 12 hours 10 minutes it was high water at Flint, and at that time there was a fall of 1 foot 7 inches to Chester; but the high water at Chester did not occur till one o'clock, by which time the water at Flint had fallen 2 feet 2 inches, and the fall on the surface of the water from Chester to Flint was 3 feet 1 inch. On referring to Plate VII., which shows the lines of ebb tide on the same day, it will be found that the water subsides gradually, and that the tidal lines approach much more nearly to parallelism and horizontality than during flood tide. The upper line of this diagram corresponds with the tidal line when it is high water at Chester.

A similar series of facts obtained on the Lune in Lancashire, will be found to corroborate generally the results deducible from those made at the Dee.

The observations at the Lune were taken at three parts of the river, namely, Glasson Dock, Heaton Point, and Lancaster Quays, the positions of which will be seen on Plate VIII. The distance from Glasson to Heaton is $3\frac{1}{4}$ miles, and that from Heaton to Lancaster $2\frac{1}{4}$ miles, making the whole distance from Glasson to Lancaster $5\frac{1}{2}$ miles.

The high-water line at Glasson, Heaton, and Lancaster, was found occasionally to stand exactly at the same height; but the average difference of level gave a fall of 1 inch from Glasson to Heaton, and a rise of 3 inches from Heaton to Lancaster, the surface of the water at Heaton being slightly depressed, and a small degree of concavity on the high-water line observable, due to the configuration of the estuary. A great contraction of the space between the banks occurs at Glasson, which checks the free flow of the tidal wave, and consequently raises its level at that place. After passing this contraction, however, the water flows into the large tidal basin or area in which the Heaton tide-gauge was placed, extending from Glasson towards Lancaster, and here the tide level again falls, owing to the much larger surface over which the water is distributed.

The time of high water was found, on eight occasions out of twenty-four, to be exactly the same at Glasson, Heaton, and Lancaster. The difference of time, however, between Glasson and Lancaster varied from 0 to 10 minutes, and the average of the observations gave the time at high water at Glasson $3\frac{3}{4}$ minutes earlier than at Lancaster; but this difference in time seemed to depend

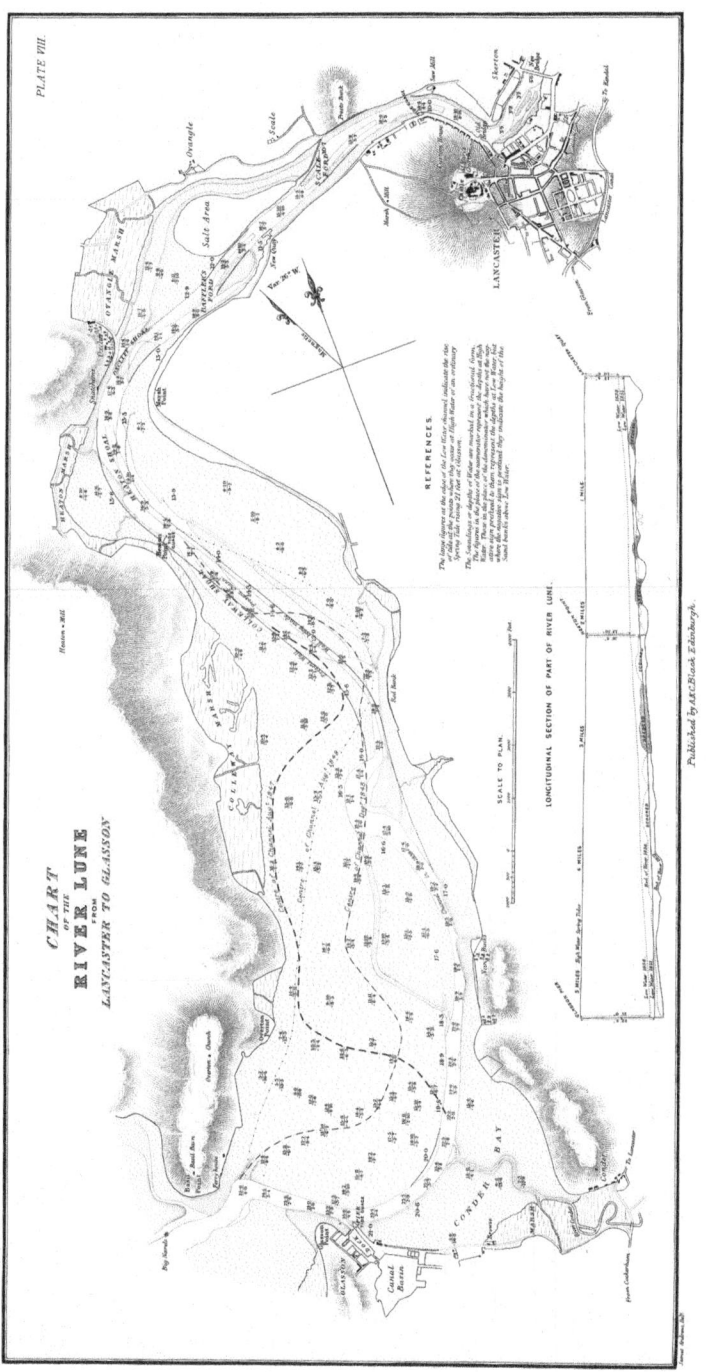

The material originally positioned here is too large for reproduction in this reissue. A PDF can be downloaded from the web address given on page iv of this book, by clicking on 'Resources Available'.

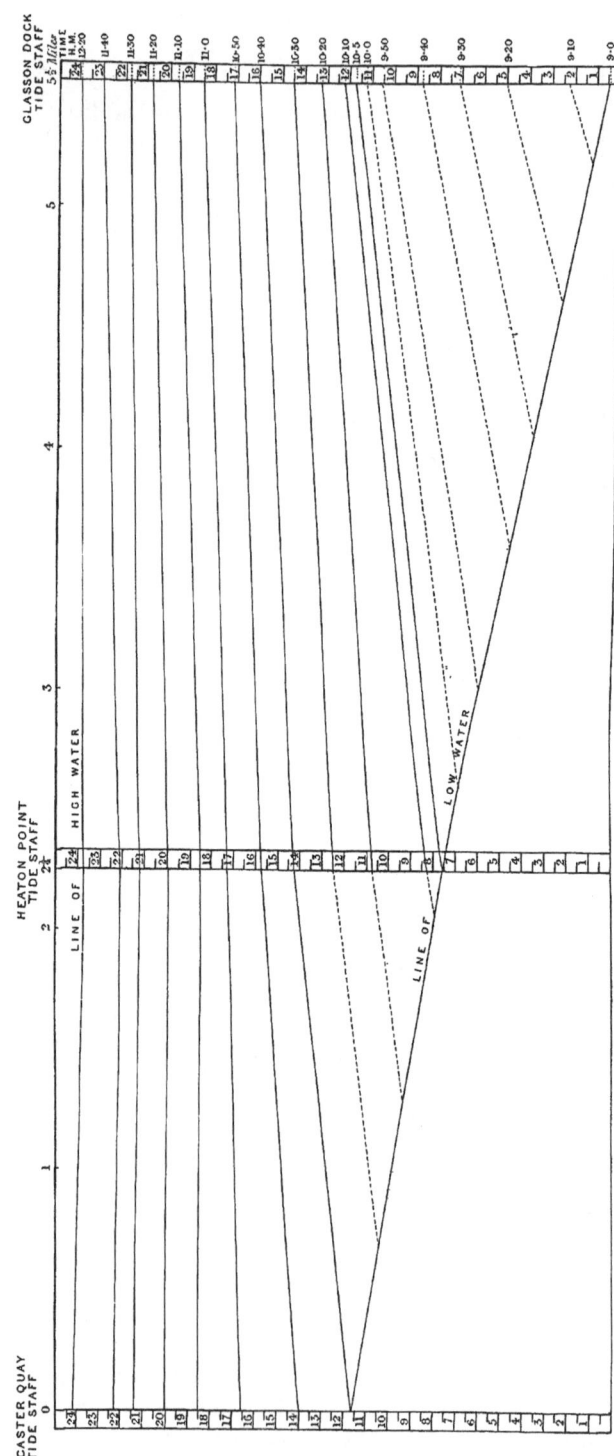

PLATE IX

RIVER LUNE.
FLOOD OF A SPRING TIDE.

Diagram shewing the approximate forms assumed by the Tidal lines of the River Lune,
During the Flood of a Spring Tide rising 23 Feet 4 Inches at Glasson.

James Andrews, Delt

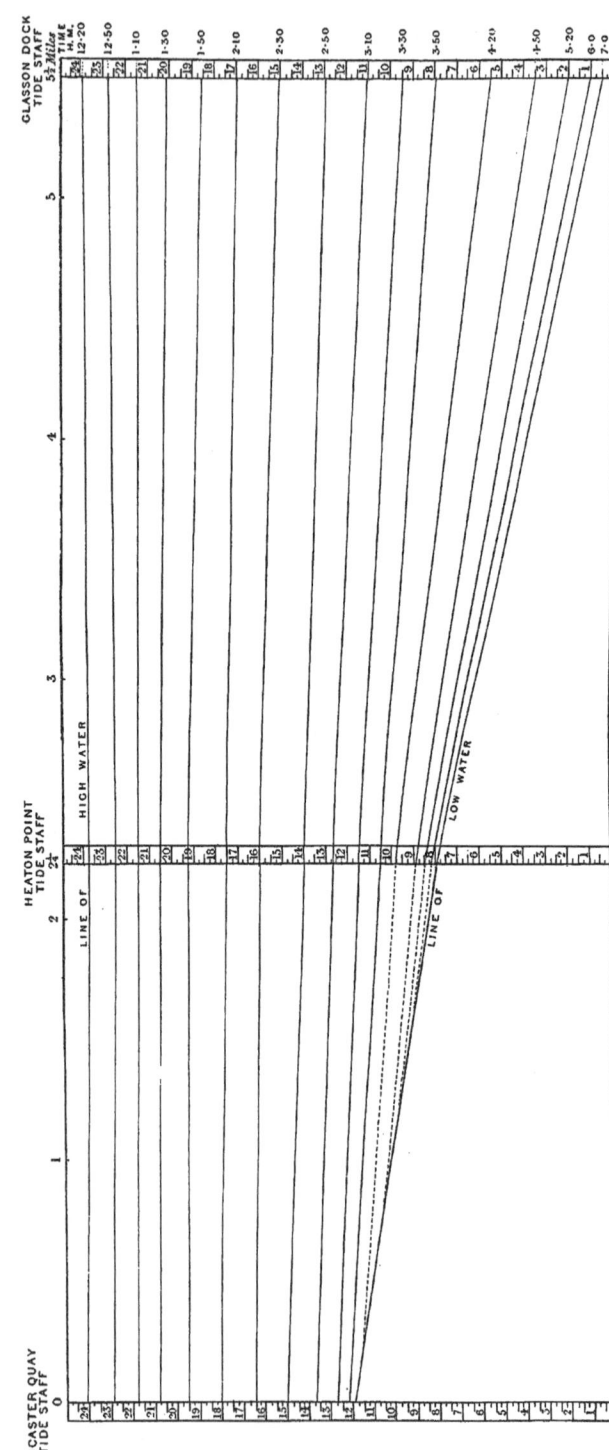

PLATE X

RIVER LUNE.
EBB OF A SPRING TIDE.

Diagram shewing the approximate forms assumed by the Tidal lines of the River Lune,
During the Ebb of a Spring Tide rising 23 Feet 4 Inches at Glasson.

entirely on the wind or state of the weather, and not on the circumstance of the tide being spring or neap.

The average level of the low water at Glasson is 7 feet 3 inches below that at Heaton, giving, in a distance of $3\frac{1}{4}$ miles, an average fall of 26·76 inches per mile; and the average level of the low water at Heaton is 3 feet 9 inches below that at Lancaster, giving, on the distance of $2\frac{1}{4}$ miles, an average fall of 20 inches per mile. The level of the low water at Glasson is, therefore, 11 feet below that at Lancaster, giving, on the whole distance of $5\frac{1}{2}$ miles, a fall on the low-water line of 24 inches per mile between the two places.

When the rise of tide, as indicated by the Liverpool tide-table, is 18 feet above the dock sill, the rise of tide in the Lune is 21 feet 1 inch at Glasson, 13 feet 10 inches at Heaton, and 10 feet 2 inches at Lancaster.

Plates IX. and X. represent the forms assumed by the tidal lines of the Lune during a spring-tide which rose 23 feet 4 inches at Glasson. The tide, as will appear from an inspection of Plate IX., began to rise at Glasson at 9 hours; at 10 hours 5 minutes it had risen 11 feet 4 inches, and at that time had just appeared at Heaton; the surface of the water at Glasson being 4 feet 3 inches above that at Heaton. At 10 hours 40 minutes the tide had risen 15 feet 6 inches at Glasson, and was 1 foot 9 inches above the level of the water at Heaton, and 4 feet 4 inches above that at Lancaster, at which place the tide had just begun to appear. Thus, while at low water there is a fall of 11 feet from Lancaster to Glasson, there was at the time mentioned a fall of 4 feet 4 inches on the

F

surface of the water from Glasson to Lancaster. At 12
hours 20 minutes it was high water at Glasson, Heaton,
and Lancaster, and at that time there was a fall of a few
inches both from Lancaster and from Glasson to the intermediate station at Heaton, producing the concavity of
the high-water line already alluded to.

On referring to Plate X., which shows the lines of ebb-
tide on the same day, it will be found that the water subsides gradually, a slight degree of concavity on the surface
being discernible for an hour and a half after high water;
and during the whole of the ebb-tide, as in the former
case, the lines approach much more nearly to parallelism
and horizontality than during flood-tide. The upper
tidal line of this diagram corresponds with that of high-
water.

I have found in all rivers whose tides I have examined
with this object in view, that, on comparing the lines
formed during spring with those formed during neap tides,
the latter are invariably more nearly parallel to the line
of high water; the deviation from parallelism decreasing
in proportion to the decrease in the rise of tide. For the
purpose of illustrating this, I have given, in Plates XI.
and XII., an example of the lines formed by the flood of
a neap tide on the Lune, and the ebb of a neap tide on
the Dee, which, when compared with the examples of the
spring-tides of these rivers already given, will be found
to approach much more nearly to horizontality and parallelism. A further illustration of this is presented in the
following tabular views of the maximum difference of
level between the surface of the water at Flint and

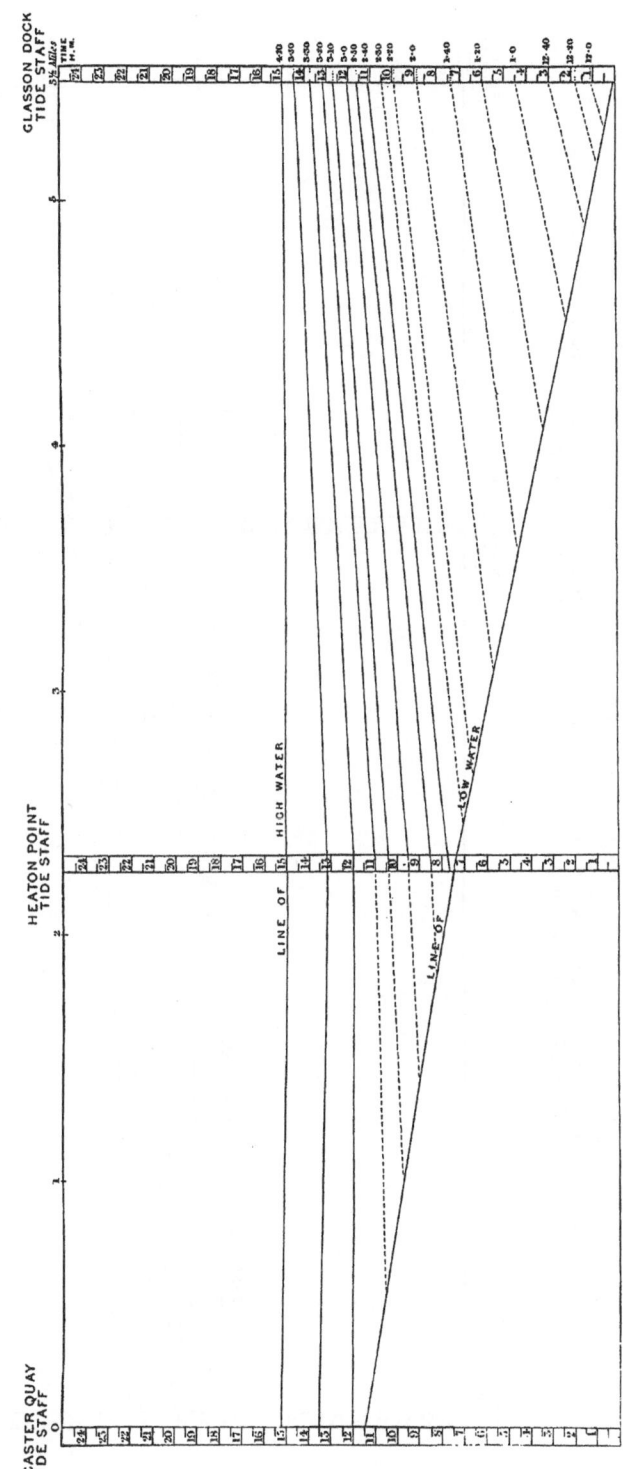

PLATE XI.

RIVER LUNE
FLOOD OF A NEAP TIDE.

Diagram showing the approximate forms assumed by the Tidal lines of the River Lune,
During the Flood of a Neap Tide rising 14 Feet 10 Inches at Glasson.

LANCASTER QUAY
TIDE STAFF

HEATON POINT
TIDE STAFF

GLASSON DOCK
TIDE STAFF

LINE OF HIGH WATER

LINE OF LOW WATER

James Andrews Del.t

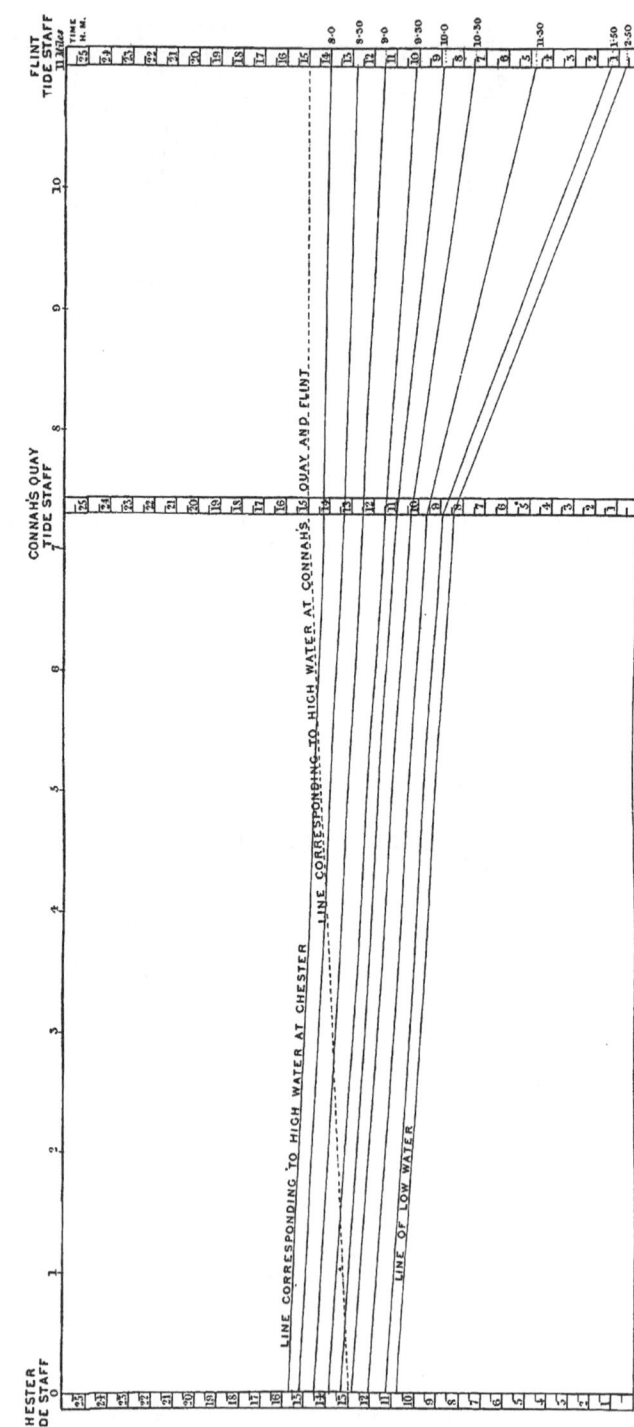

PLATE XII.

RIVER DEE.
EBB OF A NEAP TIDE.

Diagram Shewing the approximate forms assumed by the Tidal lines of the River Dee on the 22d May 1839
During the Ebb of a Neap Tide rising 4 Feet 11 Inches at Chester.

Chester on the Dee, and at Glasson and Lancaster on the Lune, during the flow of tides of various amounts of vertical rise :—

RIVER DEE.

DATE.	Rise of Tide at Flint.		Maximum Fall from Flint to Chester.	
1839.	Feet.	In.	Feet.	In.
May 21.	14	0	3	8
„ 23.	15	6	4	5
„ 25.	16	4	5	8
„ 29.	18	0	6	6
June 10.	19	8	7	10

RIVER LUNE.

DATE.	Rise of Tide at Glasson.		Maximum Fall from Glasson to Lancaster.	
1839.	Feet.	In.	Feet.	In.
Aug. 29.	12	1	1	1
„ 31.	12	9	1	6
Sept. 1.	15	4	2	0
„ 3.	19	8	2	10
„ 5.	23	2	3	2
„ 6.	23	6	4	4

I shall only refer to another example, which is chiefly interesting as showing the undulating lines of the tide-wave in its passage up the narrow channel of a winding river. I allude to the Forth in Stirlingshire, in which, from its tortuous course, the tides are somewhat remarkable. To give an idea of the windings of this river, it may be stated that the distance in a straight line between the towns of Alloa and Stirling, both of which are situated on its banks, is 5 miles; while by the river's course it is no less than 10½ miles. The tidal currents, however, in the Forth are not so rapid as those to which I have been referring, otherwise the deviations in the tidal lines would doubtless have been much greater than they were found in reality to be.

Four stations were selected, namely, at Alloa, Tillibody, Powishole, and Stirling, and the observations were made under the direction of Mr. Robert Stevenson. The deviation in the lines will be seen by reference to the diagrams in Plate XIII., which are constructed nearly in the same manner as those already described, and represent the forms assumed by the surface of the water during flood and ebb, at the end of every successive half hour. The most anomalous result of this investigation occurs at Powishole, where the undulating surface of the water was found to rise higher than at any other point on the river, either above or below it.

Although many other series of observations affording similar results might be given, it seems unnecessary to enter upon them; my only object being to enable the reader to form distinct ideas as to the nature of the deviations in the tidal lines, and the several investigations that require to be instituted in making a correct survey. The examples I have given, it is presumed, afford sufficient information for that purpose. I shall, therefore, proceed to show in what manner and to what extent, the accuracy of the soundings may be affected by the non-parallelism of the tidal lines to the line of high

PLATE XIII

RIVER FORTH.

Diagram shewing the approximate forms assumed by the Tidal lines of the River Forth.

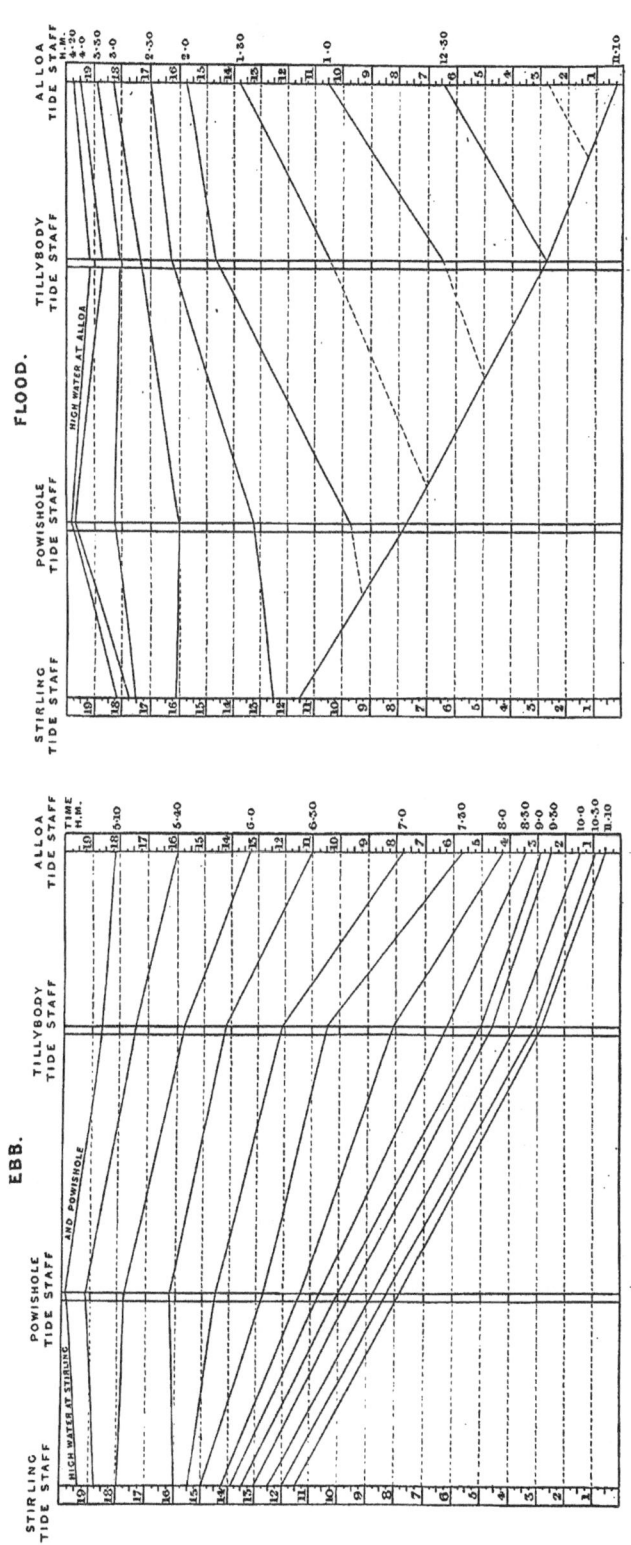

water; and, that the observations to be made may be clearly understood, I shall, in the first place, offer a few remarks on the datum to which the soundings should be reduced, and also on the nature and use of the reference which is made to the tide-gauges, in the reduction of their depths to that datum.

It is evident that all soundings must be reduced or referred to one datum line, before a correct notion can be formed of the depths of water at the places where they were taken; and perhaps the most convenient datum is the high water of an ordinary spring-tide. When I had occasion to determine this, I have taken the range of the five highest spring-tides of each series throughout the year (rejecting any *abnormal* tide due to a storm), and adopted the *mean* of these as the range of an ordinary spring-tide. This, however, requires access to a long series of observations, which is not often available, and it will frequently be found convenient to reduce the soundings to high water of a spring-tide, having a range of 16, 18, or 20 feet, as the case may be, at a certain point in the river which must be specified, and the depths in reference to the high water of any other tide, can, with this information, be easily ascertained. *Datum for soundings.*

In order to explain the use of the reference which is made to the tide-gauges, in reducing the soundings, we shall suppose that a depth was taken in the middle of an estuary, and that the observer, at the time he made the observation, had not any means of ascertaining the state of the tide. Such an observation would evidently be of no practical use, from the circumstance of its being *Use of Tide-gauges in reducing soundings.*

impossible to ascertain whether the tide had still to rise, had attained its full height, or had fallen a certain number of feet at the moment it was made, without a distinct and accurate knowledge of which, the depth could not be reduced to the level of the high water of any particular tide. If all the depths were taken exactly at the time of high water of the tide to which they were to be referred, they would not require any correction; but it is obvious that in practice this could not be done; and recourse is consequently had to observations of the rise and fall of the tide on graduated gauges, and from these the reduction is easily effected. All that is necessary for this purpose is, to note the time at which the sounding is taken, and to ascertain from the tide-gauge record, the height at which the tide stood on the nearest gauge when the sounding was made. The method of obtaining the corrected depth resolves itself into one of three cases, depending on the time of tide at which the observation was made. It is as follows :—

Formulæ for reduction of soundings.

Let a represent the depth of sounding made at a certain hour.

β the height at which the water stood on the tide-gauge at the same hour.

γ the height to which high water of ordinary spring-tides rises on the gauge.

δ the depth of the sounding reduced to high water.

Now, in the first case, if β is below the level of γ, then

$$\delta = a + (\gamma - \beta).$$

In the second case, if β is on the same level as γ, then

$$\delta = a ;$$

And in the third case, which may happen in a high spring or equinoctial tide, if β is above the level of γ, then

$$\delta = a - (\beta - \gamma).$$

These formulæ would give the true corrections of the soundings, however far removed from the tide-gauge their positions might be, if the lines formed by the tidal wave were parallel to that of high water at all times of tide, as in that case the vertical spaces $\gamma - \beta$ or $\beta - \gamma$, intercepted between the high-water line, and the other tidal lines, would be equal throughout the whole of the tidal area of the river or estuary. But it has been shown that the tidal lines are not parallel, and the formulæ I have given

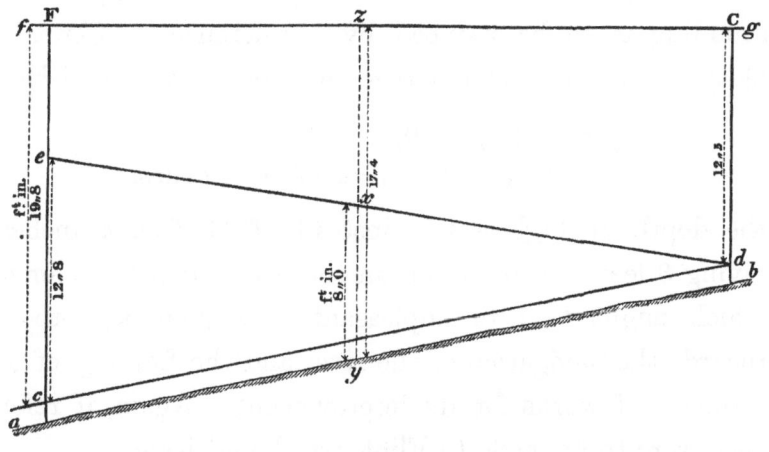

FIG. 14.

may, therefore, under certain circumstances, lead to error. As an example of this, I shall take one of the tide lines of the Dee from Plate VI.

Let F and C, fig. 14, represent the positions of Flint and Connah's Quay Tide-gauges, and the intermediate point z the place at which the sounding was taken. Let F C represent the line of high water to which it is wished to reduce the sounding, $a\,b$ the bed of the river, $c\,d$ the low-water line, and $e\,d$ the tidal line which existed when the observation was made, which is not imaginary, but will be found to correspond with that at 10 hours 15 minutes, as represented in Plate VI. Further, let the sounding $x\,y = 8$ feet. Let the depth at high water $z\,y$ at the position of the sounding, as measured on the diagram, $= 17$ feet 4 inches. Let the rise of tide at Connah's Quay $g\,d = 12$ feet 5 inches, the rise of tide at Flint $c\,f = 19$ feet 8 inches, and the height at which the water had risen on the Flint gauge, when the sounding was made, $e\,c = 12$ feet 8 inches. Now suppose the sounding is to be reduced by a reference to Connah's Quay; according to the foregoing formula we should have

$$z\,y = x\,y + (g\,d - 0)$$
$$= 8^f + (12^f.5 - 0) = 20 \text{ feet 5 inches,}$$

the depth at high water, instead of 17 feet 4 inches, giving 3 feet 1 inch more than the actual depth, an error which might lead to unpleasant consequences, both as regards the navigation of the river and the framing of an estimate of works for its improvement. Again, if reference were to be made to Flint, we should have

$$z\,y = x\,y + f\,c - e\,c$$
$$= 8^f. + (19^f.8 - 12^f.8) = 15 \text{ feet,}$$

the depth at high water, instead of 17 feet 4 inches, being an error of 2 feet 4 inches.

Now, the case that has been taken, which, in any view of the subject, would involve an error in the depth, either of 3 feet 1 inch, or 2 feet 4 inches, is not the worst that may be cited, for, under certain circumstances, and in certain situations, the error would be considerably greater.

Nor, indeed, are the high-water depths the only results that would be affected. The correctness of the section of the bed of the river, the depths of the soundings when reduced to low water, and the heights of the sand-banks above the low-water line, depends entirely on the accuracy of the high-water depths. It is obviously of great importance, therefore, that the engineer should not only be fully aware of the cause of these errors, and the extent to which the results of a survey may be affected by them, but also that he should know, and be able to apply, where necessary, the means by which they may be neutralized.

It has been shown that the erroneous results alluded to arise from the non-parallelism of the tidal lines to the line of the high water to which the soundings are to be reduced, and it has been stated that the most effectual means of avoiding inaccuracy from this cause is to increase the number of gauges ; but even this precaution, unless carried to an extent which may, in ordinary practice, be safely regarded as quite unattainable, would not produce the desired effect. The only really practicable cure which can be applied is that of taking the soundings when the lines are most nearly parallel to the line of high water. That there are not only certain tides, but also certain periods of every tide, when this approach to parallelism is much more near than at other times, has, it is presumed, been clearly established ; and in accordance with this I gave in 1842 rules for direction in making the soundings, the correctness of which I have had repeated opportunities of practically testing.

General rules for taking soundings.

First, Soundings made in the immediate vicinity of the gauge, by a reference to which they are to be corrected, are not appreciably affected by deviations from parallelism, and may be taken at any time of tide, and under any circumstances.

Second, The farther distant the positions of the soundings are from the gauge, by a reference to which they are to be corrected, the greater is the chance and the amount of error which may arise from non-parallelism.

Third, Soundings should be made during neap in preference to spring tides.

Fourth, Soundings should be made in ebb in preference to flood tides.

Fifth, Soundings to be taken in flood-tides, especially during springs, should not be made till within about an hour of high water.

If these precautionary rules be kept in view, they will be found to counteract in so great a measure the effects of non-parallelism as to insure in most cases sufficient accuracy for all practical purposes, in reducing the observations. They apply most particularly to rivers in which the rise of tide is great and the currents are strong ; but they may be said to be applicable, in a greater or less degree, to all situations, and may be embraced in these two directions, a compliance with which does not seem to involve any great difficulty : First, *soundings should not be made during very high tides;* and secondly, *all observations not made about an hour before and after high water should be confined to the immediate vicinity of the gauge by a reference to which they are to be corrected.*

In addition to the soundings which are taken throughout the area of an estuary during flood or ebb tide, it is useful to take a series in the navigable channel at low water. The soundings belonging to the first class are to be reduced to the high water of ordinary spring-tides by the formula given at page 86 ; those of the second or low-water series require no reduction, unless the level of the water on the gauge shows that the river was in flood, and when all of them have been laid down on the chart we shall have the high-water soundings distributed over the several sand-banks throughout the area of the estuary, and also a line of low-water soundings in the centre of the navigable channel, as shown in the chart of the Lune, Plate VIII., on which the soundings have been marked as an illustration. But it is still necessary to ascertain the heights of the sand-banks above low water. For this purpose, the rise of the tide at different parts of the river must be ascertained ; which, together with the determination of the height of the sand-banks above the low-water mark, is done in the following manner :—

From the system followed in taking the soundings which I have described, it is evident that a high and a low water sounding nearly corresponding in position can generally be obtained at all the places where the lines of soundings, which have been taken during flood or ebb tide, cross the low-water channel. This will be easily understood by referring to the chart of the Lune, although, from the smallness of the scale, the exact lines in which they were taken are not very definitely exhibited. Let *a*, therefore (without reference to the Plate), represent

High and low water soundings.

a high-water sounding (corrected by the formula given at page 86) whose position is in the navigable channel, and let β be the low-water sounding which most nearly coincides with the position of a. Then $a - \beta$ will be the vertical rise of tide (which we shall call γ) at that point. The values of γ being thus found at as many points as possible in the channel of the river, the number of which points will be limited by the number of the lines of high-water soundings that cross the low-water channel, they should be marked on the plan in large figures, as shown in the chart of the Lune already referred to. Now the values of the soundings $a\ a\ a$, etc., distributed throughout the estuary, will either be equal to, greater, or less than those of $\gamma\ \gamma\ \gamma$, etc., which we may suppose to represent the vertical rise of tide at the points nearest which the soundings occur, and the three results may be expressed as follows: when

Formula for ascertaining the rise of tide and height of sand-banks.

$$a - \gamma = n,$$

the sounding has occurred in the navigable channel or some deep pool in the sand-banks, and $n =$ the depth at low water; when

$$\gamma = a,$$

the sounding has been made exactly at the edge of low water, or perhaps of a pool on a level with low water, as the case may be; or, in other words, at the point where the level of low water cuts the sand-bank; and when

$$\gamma - a = n,$$

or, as it is marked on the chart, $- n$.

the sounding has occurred on a sand-bank or other raised

obstruction, and $- n =$ the height of the bank above low water. In marking these soundings on the plan it is necessary to show the depths both at high and at low water, and the most convenient way of doing this is to put them in a fractional form, the depths at high water being placed as the numerator, and those at low water as the denominator, distinguishing the heights of the sand-banks above low water by prefixing the negative sign. According to this notation the three results alluded to would be stated thus on the plan :—

$$\frac{a}{n}, \ \frac{a}{o}, \ \text{or} \ \frac{a}{-n},$$

according as the sounding had been taken in the low-water channel, at its edge, or on the top of a sand-bank.

For an example of this, Plate VIII. may be again referred to, where, at a place marked Bafflersford, nearly opposite the needle of the compass, it will be observed, by the large figures at the side of the channel, that the rise of tide is 12 feet 9 inches. It will also be seen that, at the middle of the channel, the sounding $\frac{18^{\text{ft.}} \ 9^{\text{in.}}}{6^{\text{ft.}} \ 0^{\text{in.}}}$ occurs, which denotes the depth at high water to be 18 feet 9 inches, and that at low water 6 feet, the difference between the two quantities being 12 feet 9 inches, which is the rise of tide at that place. On the adjoining sand-bank there is a sounding, $\frac{9^{\text{ft.}} \ 11^{\text{in.}}}{- \ 2^{\text{ft.}} \ 10^{\text{in.}}},$ denoting the depth at high water to be 9 feet 11 inches, and the height of the sand-bank above the level of low water to be 2 feet 10 inches.

The same notation will be perceived throughout the whole of the chart of the Lune.

Cross sections, etc.

In addition to the soundings, it is very generally necessary to make cross sections and borings of the river's bed to ascertain the quantities and quality of the materials to be excavated in order to obtain a certain depth of water.

In selecting the points at which to make these observations the engineer must of course be guided by the object of the investigation and the formation of the river's course.   If there be fords or shoals in the bottom which occasion obstructions to the navigation, and require to be removed, one or more lines of section may be fixed on at each shoal, according to its extent.   Where, as sometimes happens, the channel is irregular, or has rock occurring at various points, it is often necessary to obtain, by means of numerous cross sections, an exact survey of the whole, or at least of a great part of the bed, before any distinct plan of operations can be formed; and the depths of the sections and borings may be referred to the same datum as the soundings of the depths of water.

Mr. Mitchell's rule for determining elevations along a tidal river without levelling.

Mr. Henry Mitchell, of the United States Coast Survey, who has published several interesting papers on Marine Surveying, has proposed an ingenious method of determining elevations along the course of a tidal river without the aid of a levelling instrument, which he explains as follows:—" Set up graduated staves at such distances apart that the slacks of the tidal currents shall extend from one to another.   By simultaneous observations ascertain the difference in the readings of those gauges at the slack between ebb and flood currents, and again the difference at the slack between flood and ebb, then apply the rule : *The difference in the elevations of the zeros of the gauges is equal to one half the sum of the differences of their readings at the two slack waters.*

"In the Hudson I found that staves 10 miles apart could be referred to each other by this rule, and that no nice current observations were really necessary. The slope is so nearly constant about the time of slack water that an error in this time of a half hour, in some cases, would be of no consequence. The *coincidence of time* at the two gauges and the careful reading of the heights are the most important elements. I offer below an illustration from observations upon one of the fourteen reaches examined during October 1871 :—

Slack, Ebb to Flood.				Slack, Flood to Ebb.			
Time.	Heights at Barnegat.	Heights at Poughkeepsie.	Difference.	Time.	Heights at Barnegat.	Heights at Poughkeepsie.	Difference.
H. M.	Feet.	Feet.	Feet.	H. M.	Feet.	Feet.	Feet.
10 20	2·07	2·86	0·79	4 45	4·24	5·73	1·49
10 25	2·18	2·97	0·79	4 50	4·20	5·66	1·46
10 30	2·27	3·07	0·80	4 55	4·12	5·60	1·48
10 35	2·34	3·17	0·83	5 00	4·03	5·53	1·50
10 40	2·47	3·27	0·80	5 05	3·99	5·48	1·49
10 45	2·54	3·36	0·82	5 10	3·93	5·40	1·47
		$a=$	0·80	5 15	3·88	5·35	1·47
				5 20	3·88	5·29	1·49
						$b=$	1·48

$\frac{1}{2}(a+b) = 1\cdot14 =$ difference of elevation of zeros of staves.
$\frac{1}{2}(b-a) = 0\cdot34 =$ slope of surface at slack water."

It is obvious that this method of treatment assumes

that the slope on the surface of the river during the slack of the *ebb*-tide is exactly equal in the opposite direction to the slope during the slack of the *flood*-tide ; that the section lines formed by the flowing and ebbing tide at these periods are extremely regular, that the gradients between the points of observation are uniform, and that the tidal lines during the period of observation are practically parallel. In the river Hudson the tidal range, though only about $5\frac{1}{2}$ feet at New York, is felt for a distance of 150 miles, as far as Albany, and the tidal lines have gentle gradients, and in that case Mr. Mitchell has found the method to be applicable ; but a glance at the tide lines in the preceding Plates will show that it could not be applied to such rivers as I have been speaking of, where I suspect nothing short of actual levelling of the gauges will insure such a result as can be relied on.

CHAPTER V.

IT is often necessary in the practice of engineering to determine the discharge of rivers, and the velocity and direction of surface and under currents. In some investigations, also, it is desirable to ascertain the quality of water taken from various depths and at different times of tide, so as to know the proportions of sea and fresh water which constitute the mixture, and the quantity of solid materials held in mechanical suspension, such as sand or mud.

A few remarks on the mode of conducting these different investigations will form the subject of this chapter.

The most accurate method of measuring water discharge is to construct a gauge-weir and ascertain the quantity of water flowing over it; but as this process is

only applicable to small streams, it does not come within
the scope of our subject.

Discharge of
rivers.

The discharge of a river is generally ascertained by
multiplying its mean velocity by its sectional area; and
in gauging a river with this object in view, it is necessary,
first, to determine accurately its sectional area in a plane,
as nearly as possible at right angles to the direction of the
current. This is done by selecting a place where the
banks are regular and the stream tranquil. A graduated
cord is stretched across, as nearly as possible, at right
angles to the direction of the currents. The depths of
water are carefully taken, with a rod graduated to feet
and inches, at distances of 5 or 10 feet (as indicated by
marks on the cord), according to the minuteness of the
inquiry to be instituted or the irregularities of the river's
bed, and, from the data thus obtained, an accurate cross
section showing the sectional area of the river can be
constructed.

Mode of deter-
mining the velo-
city of a river.

After the observations for the cross sections have been
completed, the measurements for ascertaining the velocity
should at once be made before any change in the level of
the water, and consequent change in the area, can take
place. The velocity with which the water passes over
the bed of the river will be found to vary, gradually
decreasing from the fair-way or deepest part of the river
towards the sides, and from the surface towards the bottom,
except in certain exceptional cases, to be afterwards
noticed. For the purpose of calculation, therefore, the
mean velocity must be determined. This is most accurately done by ascertaining the surface velocity in the

middle of each of the compartments into which the transverse section of the river is divided, by the soundings made, as already explained, and from these surface velocities, by a simple formula, the mean velocity of each of the compartments can be obtained, and the mean of these will be the required *mean* velocity of the river.

For the purpose of ascertaining the surface velocities, various methods may be employed.

The most common, but by no means the most satisfactory, mode of proceeding, is to drop into the water, from a boat, a float (whose specific gravity is merely great enough to sink it to a level with the surface), at a point about 30 or 40 feet above the line of section, so as to insure its acquiring the full velocity of the current before it reaches the cord. An observer, stationed at the cord, notes exactly the moment at which the float passes, and follows it down the stream till he reaches the line of two poles, which have been fixed in reference to the observations, when he again notes the exact moment of its transit at the lower station. The elapsed time between the two transits is then noted in the book, along with the distance between the two places of observation, which, owing to the irregularity of most rivers, with regard to width, depth, and velocity, can seldom be got to exceed 100 feet. This operation has, of course, to be repeated for every compartment of the cross section. By floats.

Dr. Anderson, in measuring the discharge of the Tay, at Perth, used an adjustable float, which extended from the surface to near the bottom of the river, and so obtained at once, approximately, the *mean* velocity of each compartment of the cross section of the stream.

Certain disadvantages attend the method of measuring velocities by floats, which render it not generally applicable. For example, it is only adapted to rivers of limited breadth, owing to the impossibility of an observer being able to discover with sufficient accuracy the exact time when the float passes the station lines, if it be viewed from a distance, as, for example, from the bank of a broad river. There are, however, greater objections than this, which, when pointed out, will be sufficiently obvious to every one. In any part of the river's bed passed over by the floats, the slightest irregularity of the bottom produces a disturbance in the motion of the stream, and alters the velocity, so that it is not possible, from the time occupied by the passage of the float over the measured distance, to deduce the mean velocity at the line of cross section. It is also impossible, by this method, to obtain a sufficient number of distinct and independent observations, applicable to each division of the stream, as the eddies and irregularities of the current which exist in all rivers, generally cause the lines passed over by the floats to cross and interfere with each other in such a manner as to destroy all connexion between any given series of observations, and the several compartments of the river, whose mean velocity they were intended to ascertain.

By tachometer. The great object is to determine the velocity of each portion of the stream, as it passes the line of cross section ; and the best way of doing this is to employ the tachometer or stream-gauge—an instrument of great service in such inquiries. The current impinging on a vane causes it to revolve, and the number of revolutions made by the vane being registered on an index, which is acted on by a set of toothed wheels, indicates the velocity of the current.

The construction of this instrument, and the manner Description of in which it acts, will be best described by a reference tachometer.

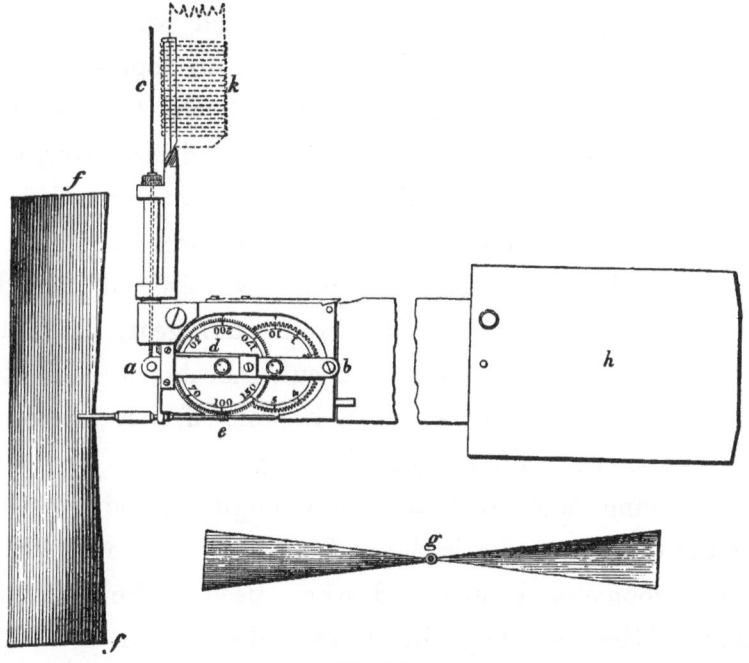

FIG. 15.

to fig. 15, which is drawn to a scale of one-third of the full size. In this view, *f f* represents the driving vane, which is acted on by the stream, and of which *g* is a plan. The plane of this vane is twisted, as represented by the dark shading in the cut, so as to present, not a knife-edge, but an oblique face to the action of the current, which, by impinging on it, causes it to revolve. On the spindle or shaft of this vane, an endless screw is fixed at *e*, which works in the teeth of the first registering wheel, and causes it to revolve, when the vane is in motion, and the screw in gear. Letters *a* and *b* represent a bar of brass, to which the pivots on which the registering wheels revolve are attached. This bar is moveable on a joint at *b*; and at the point *a*, a cord, *a c*, is fixed, by pulling which the bar and wheels can be raised, and on releasing it they are again depressed by a spring at *d*. When the bar is raised, the teeth of the wheel are taken out of gear with the endless screw, and the vane is then left at liberty to revolve, the number of its revolutions being unregistered; but when the cord is released, the spring forces down the wheels, and immediately puts the registering train into gear, in which state it is represented in the cut. Letter *h* is a stationary vane (which is shown broken off, but measures about 9 inches in length) for keeping the plane in which the driving vane revolves at right angles to the direction of the current, and *k* is the end of a wooden rod to which the tachometer is attached when used. The different parts of the instrument itself are made of brass.

The moveable bar for the registering wheels and the application of the cord and spring which have been described, afford the means of observing with great accuracy in the following manner. The instrument having been adjusted by setting the registering wheels at zero, or noting in the field-book the figure at which they stand, the cord is pulled tight, so as to raise them out of gear, and the instrument is then immersed in the water. The vane immediately begins to revolve by the action of the current, and is permitted to move freely round until it has attained the full velocity due to the stream, when a signal is given by the person who observes the time, and the registering wheels are at that moment thrown into gear by letting the cord slip. At the end of a minute another signal is given, when the cord is again drawn and the wheels taken out of gear, and on raising the instrument from the water, the number of revolutions in the elapsed time is read off. This observation being made in the centre of each division of the cross section, the number of revolutions due to the velocity at each part of the very line where the cross section is taken is at once obtained.

Before using the tachometer, it is obvious that the value of a revolution of the vane must be ascertained; and although this is done by the manufacturers, it is proper that the scale of each instrument should be determined by the person who uses it, and that it be tested if the instrument has been out of use for some time, before being again employed in making observations. A scale sufficiently accurate for most hydrometric purposes may be obtained by immersing the instrument in some regular channel, such as a mill-lead formed of masonry, timber, or iron, where the velocity is nearly the same throughout, and noting the number of revolutions performed during the passage of a float over a given number of feet, measured on the bank. This number, therefore, becomes a constant multiplier, and the number

of revolutions being determined, the number of feet
passed over by the water in the given interval of time
is ascertained.

The values of the revolutions may perhaps be more
accurately determined by placing two stakes 100 feet
apart on the banks of a canal. The gauge, attached to
a rod having a knee or bend at its extremity, is then
immersed at a little distance from the stakes, and drawn
quickly through the water, so as to cause the vane to
revolve. On passing the first stake the cord is slipped,
and the registration commences. On passing the second
stake the vane is taken out of gear, and the number of
revolutions made in passing over the distance of 100
feet gives their value. The operation is repeated several
times, alternating the direction in which the gauge is
moved through water, to destroy the effect of any small
current that may possibly exist, and the mean of the
observations is adopted to calculate the scale of the instrument.

Formula for
reducing the
surface to mean
velocity.

Having thus by means of the tachometer determined
the surface velocity of the river at each of the divisions
of the extended cord, the next step is the reduction of the
observed surface to those of mean velocities, which will
be readily done by the following rule of Du Buat.

It is not clear that Du Buat meant this formula to be
applied in the manner here described; on the contrary, it
rather appears that he meant to deduce from a single
surface velocity taken in the centre of the stream a mean
velocity applicable to the whole sectional area. He, as
quoted by Professor Robison, says, " The mean velocity

in any pipe or open stream is the arithmetical mean between the velocity in the *axis* and the velocity at the sides of a pipe or bottom of an open stream;" but it is hardly possible to find a river with a cross section so symmetrical as to admit of a single central observation proving sufficient, and the formula is therefore often applied to the velocities as measured in the centres of the different compartments into which the river is divided in making the cross section. Du Buat's rule referred to is as follows :—

If unity be taken from the square root of the surface velocity expressed in inches per second, the square of the remainder is the velocity at the bottom, and the mean velocity is the half sum of these two.

Thus, let a = the observed surface velocity,

,, β = the bottom velocity, and

,, γ = the mean velocity, all in inches per second.

$$\beta = (\sqrt{a} - 1)^2 \text{ and } \gamma = \frac{a + \beta}{2} ;$$

and hence, the mean velocity is directly deducible from the surface velocity by the following formula :—

$$\gamma = \frac{a + (\sqrt{a} - 1)^2}{2}.$$

The mean velocities obtained by calculation are to be multiplied into the area of the spaces in the centres of which the observations were made, in order to obtain the cubic contents of water discharged in each division ; and to obtain the whole discharge, it is only necessary to add together the results of the observations made in all the different compartments. The apportioning of the stream into different parts, and treating each as a separate channel, appears to insure a much greater probability of a correct measurement than any method which depends upon assigning to the whole area a common velocity ; and it is obvious that this method can be effectually followed only by the use of the tachometer described, or by any similar instrument which possesses the advantage of confining its indications to the spot where the sectional area of the river is *actually* measured. Wherever, as will frequently happen in regular streams, the velocity of several compartments, as ascertained by the stream-gauge, are found to be the same, the areas of these compartments may be added into one sum and multiplied by the common velocity. It seems necessary to observe that velocities exceeding 3 miles an hour are apt to injure an instrument of the size and proportions shown in the cut, and that in gauging more rapid rivers an instrument on the same principle, but of stronger make, should be employed.

The velocity of currents in the open sea or in estuaries may also be determined from a boat at anchor, by allowing a float to run out during a given interval of time, and observing the quantity of graduated line which has been let out. Some of the various forms of registering logs are also very suitable for such experiments when the velocities are not below 2 miles an hour.

Methods of ascertaining discharge by formulæ.

But as the numerous observations of velocities which I have described always occupy much time, many formulæ have been proposed to shorten the work of calculating the discharge of a stream. I have had opportunities of testing the value of these formulæ by comparing the results they gave with those obtained by the more careful and elaborate process which I have described, and as these generally recognised formulæ are conflicting, and in some cases inaccurate, I shall give the result for the information of the student.

Before doing so, however, it is necessary to define and explain certain terms, without which the application of the different formulæ would not be intelligible.

In dealing with the discharge of a river, we are to understand :—

First, That the *slope* is the fall on the surface of the water, and is generally expressed in feet per mile, and is ascertained by levels carefully taken.

Second, The *sectional area* is the width multiplied by its average depth, as ascertained by means of the section already explained at page 98.

Third, The *hydraulic mean depth* is the quotient given by dividing the sectional area of the channel in square feet by the wetted border or perimeter in lineal feet, also ascertained from the section.

Fourth, The *mean velocity*, which may either be deducted from the surface velocity by formulæ, or ascertained directly by measurement, is that velocity which is used in ascertaining the discharge.

Fifth, The *discharge* is the quantity of water yielded by the stream in a given time, and is generally stated in cubic feet per minute, being the mean velocity in feet per minute multiplied by the sectional area in square feet.

The formulæ which I subjected to trial were :—

I. Formula given by Dr. Robison, founded on Du Buat's investigations :[1]—

$$M = \frac{307 \; (\sqrt{\overline{d}} - 0\cdot1)}{\sqrt{S} - \text{Hyp. log. of } \sqrt{S} + 1\cdot6} - 0\cdot3 \; (\sqrt{\overline{d}} - 0\cdot1)$$

in which M = the mean velocity in inches per second,

　　　　d = the hydraulic mean depth in inches,

　　　　S = the reciprocal of the slope of the surface which is the denominator of the fraction expressing the slope, the numerator being always unity (a slope of 1 foot a mile is $\frac{1}{5280}$, therefore 5280 = reciprocal for that slope),

Hyp. log. = the common log. of the number to which it is attached, multiplied by 2·3026.

II. Formula given by Sir John Leslie :[2]—

$$M = \frac{15}{16} \sqrt{\overline{af}}$$

in which M = the mean velocity in miles per hour,

　　　　a = the hydraulic mean depth in feet,

　　　　f = the fall on the surface in feet per mile.

III. Formula given by Mr. Ellet for calculating discharge of the Mississippi :[3]—

$$V = \frac{8}{10} \sqrt{\overline{df}} + \frac{df}{20}$$

$$M = 0\cdot8 \; V$$

in which V = the surface velocity in feet per second,

　　　　d = the maximum depth of the river in feet,

　　　　f = the fall on the surface in feet per mile,

　　　　M = the mean velocity in feet per second.

---

[1] See article " River," *Encyclopædia Britannica ;* also *A System of Mechanical Philosophy,* by John Robison, vol. ii. p. 453.

[2] *Elements of Natural Philosophy,* by Prof. Leslie, Edinburgh, 1829, vol. i. p. 423.

[3] *The Mississippi and Ohio Rivers,* by Charles Ellet, Philadelphia, 1853.

IV. Formula given in Mr. Beardmore's tables :[1]—

$$M = \sqrt{a}\,2f \times 55$$

in which M = mean velocity in feet per minute,

    a = hydraulic mean depth in feet,

    f = fall per mile in feet.

V. In addition to these formulæ, the writer also subjected to trial the formula of Du Buat, as given at page 105 :—

$$M = \frac{(\sqrt{V} - 1)^2 + V}{2}$$

in which M = the mean velocity in inches per second,

    V = the maximum surface velocity in the axis of the stream in inches per second.

In order to compare these different formulæ, a very favourable situation was selected for ascertaining the discharge of a stream by careful measurements of its sectional area and of the velocities at different parts of its surface from the centre to either side with the tachometer, as described at page 102, and the result gave a discharge of 1653 cubic feet per minute, which, from various measurements, I believe to be a very near approximation to the actual discharge. The slope was also accurately ascertained by careful levellings, and the following are the results :—

                               Cubic feet.

Discharge from measurement as above, 1653 per minute.

1st, By Robison's formula,   .   . 2214   do.

2d, By Leslie's   do. .   .   . 2474   do.

3d, By Ellet's   do. .   .   . 2784   do.

4th, By Beardmore's do. .   .   . 2335   do.

5th, By formula assuming the mean deduced from the centre surface velocity as the mean for the whole section, . 1950   do.

---

[1] *Hydraulic Tables*, by Nathaniel Beardmore, C.E., London, 1852.

It will be seen from this statement, that none of the formulæ afford a near approximation to the discharge of the small stream to which they were applied.

Again, it was ascertained by the late Dr. Anderson, of Perth, after most carefully dividing the cross section into compartments, and ascertaining the velocity of the stream in each of them, by the method described at page 99, that the discharge of the main branch of the Tay at Perth was 147,391 cubic feet per minute.[1] I ascertained the discharges, as calculated by the different formulæ as above, and the following are the results :—

Cubic feet.

Discharge per measurement, by Dr.
　　Anderson,　.　　.　　.　　.　　. 147,391 per minute.
1st, By Robison's formula,　　.　　. 153,632　　do.
2d, By Leslie's　　do.　.　　.　　. 166,134　　do.
3d, By Ellet's　　do.　.　　.　　. 122,002　　do.
4th, By formula in Beardmore's tables, 156,569　　do.
5th, By formula assuming the mean de-
　　duced from the centre surface velocity
　　as the mean for the whole section, . 179,237　　do.

Formula gener-
ally applicable,
but affording
only an ap-
proximation.

The result of these trials, and others which I have had occasion to make, is, that none of the formulæ that have been proposed will be found *generally* applicable ; but the following formula may be applied, and will, in most cases, give a pretty near approximation to the velocity and discharge due to a given area and fall, viz. :—

$$x = y \sqrt{af}$$
$$z = \frac{x \times 5280}{60}$$
$$D = sz$$

---

[1] This does not include the Willowgate nor the Earn.

in which $x =$ the mean velocity of the whole section of the stream in miles per hour,

$y =$ a quotient which is found to vary from 0·65 for small streams under 2000 cubic feet per minute, to 0·9 for large rivers, such as the Clyde or the Tay,

$a =$ the hydraulic mean depth in feet,

$f =$ the fall on the surface in feet per mile,

$z =$ the mean velocity of the whole section of the stream in feet per minute,

$s =$ the sectional area of the stream in feet; and

$D =$ the discharge in cubic feet per minute.

It must still be kept in view that the application of any known formula to the determination of the mean velocity and discharge of a river is shown, by experimental inquiry, to afford only a rough approximation; unless observations are made embracing the velocities at different parts of the cross sectional area, in the manner already described at page 98.

In order to render the measurement of discharge Floods. useful, care should be taken, when the stream is gauged, to ascertain that it is in a *normal* condition, by which is meant that it is neither dried to its *minimum* by a long drought, or swollen to its *maximum* by heavy rains. The stream in this *normal* condition is said to be in its state of *ordinary summer water*, or at its ordinary *summer water level*, and to be unaffected by long droughts or by heavy rainfall.

It is obvious that it is not possible to offer any directions for determining when a stream is in this *normal* condition, but it will generally be found that the residents on its banks, particularly those engaged in its

fishings, if there be any, can tell when the water is at
ordinary summer level. The fluctuations of a river from
its lowest to its highest state are excessively capricious,
the amount of flooding which is ascertained to take place
in different rivers having no constant ratio either to the
summer water which they discharge or to the area
drained by them. This, indeed, does not seem surprising
when we consider the very different character, both geolo-
gically and agriculturally, of the districts through which
rivers flow. The drainage area in one situation may
include large tracts of hill country, having steep and
scantily soiled slopes, from which the rain is readily discharged; in another place it may be flat, or gently rising
deep soiled agricultural land absorbing much of the rain
that falls, and giving it off only by slow degrees. Other
districts are more or less affected by their geological formation—some strata being less absorbent than others. In
others, again, agricultural improvements have an influence
on the drainage—sheep-grazing land being less absorbent
than arable land. In rivers which flow from lakes a
reservoir is afforded for the storage of surplus water,
which checks the floods below. But again, as in the case
of the Tay, which flows from Loch Tay, a sheet of water
fourteen miles long and three-quarters of a mile broad, it
is found that in gales of westerly wind accompanied by
heavy rain the lake water is heaped up at the outlet, and
greatly increases the flood in the river; so that even in
the recurrence of floods themselves there are many circumstances which vary their effects, even in the same
district. The heaviest floods in all rivers occur with

heavy rain and melting snow, for then the bed of the river has, it may be said, to discharge a compound flood made up of melting snow and falling rain.

The construction of railways in India has afforded interesting information as to the floods of the great Indian rivers, which are fully discussed in papers by Lieutenant-Colonel O'Connel,[1] and Mr. Howden.[2] The consideration of the data thus obtained has suggested various formulæ for calculating the discharge due to a given area, but the information as to the amount of flood water said to have been discharged from different districts of country is so discordant, that it seems to me to be impossible with elements so variable to found any formula that can be generally useful.

The quantity passing off during high floods is variously stated by different authorities from 1 foot to 30 cubic feet per minute per acre according to the district in which the observations were made.[3] But the highest gauging I have ever got was 15 cubic feet per acre from a town district of 630 acres, after three days of nearly continuous rainfall. Thunderstorms discharge a very much greater amount during their short duration. It is stated that in August 1846, during a thunderstorm, 3·3 inches fell in 2 hours and 20 minutes, being 85 cubic feet per minute per acre.[4]

Perhaps the only general result to be gathered from

[1] *Minutes of Proceedings of Institution of Civil Engineers*, vol. xxvii. p. 204.

[2] *Ibid.* p. 218.

[3] *Ibid.* vol. xxi. p. 84. "Rainfall and Evaporation," by A. Leslie, C.E., *Trans. Royal Scot. Soc. of Arts*, vol. viii.

[4] Parliamentary Report on Metropolitan Main Drainage, 1858.

the published observations relative to floods, is that *the flood discharge has a higher ratio to the ordinary discharge in small than in large rivers.* This is due very much to the fact that in a small river a rain-fall affects every one of its feeders, whereas in a larger river the influence of the rain is limited to one portion of the district only. If, for example, such a river as the Mississippi were subjected to an increase of its bulk similar to that of small rivers, the country through which it flows would be entirely devastated. The safety of such a country is due to the important fact that excessive falls of rain, like hurricanes of wind, while at the height of their fury are not wide spread, but act on a comparatively limited portion of the earth's surface.

Though we cannot, therefore, deduce from data so arbitrary any law applicable to rivers in all districts, we are not precluded from dealing with the different sizes of floods discharged from any particular district; and as it is sometimes desirable to ascertain by gauging the *average summer* flow of a stream, I give the following mode of computing the discharge, *exclusive of floods*, which has been proposed by Mr. Leslie.[1] "First, the gaugings are all to be set down in a table in the order of their quantities, beginning at the smallest and going on to the largest, or *vice versa.* The whole number of observations is then to be divided as nearly as possible into four equal parts; whereof the lowest fourth is held to comprehend the *extreme droughts*, and the highest the *floods*.

Mr. Leslie's method of gauging average discharge, exclusive of floods.

[1] *Minutes of Proceedings of Institution of Civil Engineers*, vol. x. p. 327.

The average of the middle half is to be ascertained, and all above that quantity is held to be flood-water.

" A new table is then to be constructed, in which all the gaugings not exceeding the average of the middle half are put down at their actual quantity; but all above the average are put down as equal to that average quantity. The average of the whole of the new table is to be considered as being a fair estimate of the water flowing in the stream, *exclusive of floods.*"

UNDER CURRENTS.

I must offer the further caution, that those rules, from which the mean velocity is deduced, on the assumption that it bears a constant ratio to the surface velocity, do not apply in many situations which are within the influence of the tide. In surveying the Dee at Aberdeen in 1812, for example, Mr. Robert Stevenson found that, while there was an *outward* upper current of fresh water, there was an *inward* under-current of salt water; so that, although the upper stratum was constantly running towards the sea, there was a regular rise and fall of the surface, produced by the influx of the tidal waters below. Another instance of such an under-current, though not occasioned by the presence of a river, was found to exist in a marked degree at the Cromarty Firth, where Mr. Alan Stevenson, in 1837, found currents greatly exceeding the surface velocity.

Results of formulæ destroyed where under-currents exist.

It is essential in some inquiries to ascertain to what depth the currents penetrate, and whether under-currents exhibit the same phenomena in regard to direction and

Instruments for ascertaining under-currents.

velocity as those of the surface; and as these inquiries are interesting and important, and have lately been much discussed in connexion with deep-sea researches, they are worthy of detailed notice.

Tachometer.   For small depths the tachometer of Woltmann, which has been already described, is a convenient and accurate instrument for measuring under-currents.   I never used it myself for depths exceeding a few feet, but I understood from Professor Gordon that it has been employed in Germany for measuring velocities at great depth, by using an apparatus erected on a platform, supported on two boats, and that Raucourt used it to measure the velocity of the Neva at St. Petersburg, at depths of 60 feet; Defontaine the Rhine, at upwards of 40 feet; and Funk many rivers, at depths of from 40 to 60 feet.   But as its application under such circumstances may be regarded rather as a purely scientific than as an engineering experiment, it is not necessary to describe it in this place. The *direction* of the under-currents, which it is sometimes interesting to know, cannot, however, be obtained by means of the tachometer, and I shall describe the plan for obtaining an *approximation* to both the velocity and direction of under-currents, which was devised and used, I believe, for the first time, at the Cromarty Firth, in 1837, by the late Mr. Alan Stevenson, when he detected the tidal anomalies already alluded to.   It may be well to explain that the waters of the Cromarty Firth pass to and from the sea through the narrow gorge between the Suters of Cromarty, where the width is about 4500 feet, and the depth about 150 feet.   The mean velocity due to the column of water passing this gorge, as deduced from the observed surface velocity, was not sufficient to account for the quantity of water actually passed during each tide, as determined by measuring the cubical capacity of the basin of the firth. This led to the observation of the under-currents through the gorge by means of submerged floats, and it was found that during flood-tides the surface velocity was 1·8 mile per hour; while at the depth of 50 feet the velocity was not less than 4 miles per hour, being an increase of 2·2 miles per hour. During ebb-tide the surface velocity was 2·7 miles per hour, and at 50 feet it was not less than 4·5 miles per hour, being an increase of 1·8 mile per hour. The instrument by which these velocities were measured consisted, as shown in figure 16, at letter *a*, of a flat plate of sheet-iron,

Under-current
float used at
Cromarty Firth

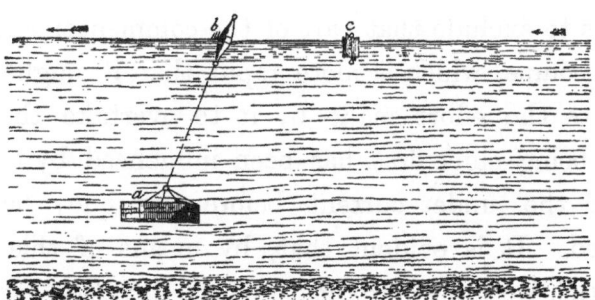

FIG. 16.

measuring 12 by 18 inches, having a vane made of the same material, and measuring 4 feet in length, fixed at right angles to the centre of it. The lower edges of the plate and vane were loaded with bars of iron, for the purpose of causing the instrument to sink to the requisite depth; and it was so slung by the cords suspending it as to preserve the surface of the plate in a vertical plane.   This apparatus was secured by a cord of sufficient length to sink it to the required depth, and the whole was attached to a tin buoy, letter *b*, which floated on the surface, its form being such as to produce little resistance to its passage through the water.   The buoy served not only to preserve the vane plate at the same depth, but also indicated its progress through the water in a very satisfactory and often interesting manner.

The plate, sunk at the depth of 50 feet, when acted upon by the force of a strong under-current, was hurried along, carrying the buoy, which floated on the surface, along with it, as shown by the buoy passing the floats thrown out on the water as gauges of the velocity of the upper current, one of which is shown at *c*.   The only precaution to be observed in making such observations, is to exclude that part of the commencement of the buoy's course, which is more rapid than it ought to be, owing to the effort made by it to overtake the plate, which, being sunk first, has been influenced by the velocity of the under-current before the buoy has been launched.   It is evident that, by means of this simple apparatus, we can approximate to the direction as well as to the velocity of under-currents; but it must be kept in view that there are several deranging influences in operation, which tend to render the results obtained merely approximations to the truth.

Since I first described these Cromarty Firth observations in 1842, many efforts have been made to ascertain the existence and strength of under-currents.

Messrs. Carpenter and Jeffreys, in 1870, when engaged in their deep-sea researches in the " Porcupine " surveying ship, endeavoured to ascertain the state of the under-currents at the Straits of Gibraltar.[1] The apparatus adopted by them for this purpose was arranged by Captain Calver, and was identical *in principle* with that employed at the Cromarty Firth; the only difference being that the under-current float was composed of a basket, with pieces of sail-cloth fixed to it, and so disposed as to catch the current. The float was weighted with lead, and the cord by which it was suspended, instead of being attached to a float as at the Cromarty Firth, was fixed to a boat, the drifting of which indicated the force and direction of the under-current. It does not appear that more than one or two observations were made with this instrument.

Under-current float used in deep-sea researches.

Captain Spratt, who has made several observations on the under-currents of the Sea of Marmora and the Dardanelles, in a paper on the under-current theory of the ocean, in the *Proceedings of the Royal Society*,[2] states the following as the plan he adopted:—" I never attempted such experiments by the use of any bulky object, such as a boat that offered great resistance to the surface current. I felt too that a fixed object, as a point of reference, was always necessary, such as a buoy or float attached to a sinker actually on the bottom. Such observations for testing ocean currents should only be made in connexion with a fixed object attached to the

Under-current float used by Captain Spratt

---

[1] *Report on Deep-sea Researches*, in July, August, and September 1870, by W. B. Carpenter, M.D., F.R.S., and J. Gwyn Jeffreys, F.R.S.

[2] *Proceedings of the Royal Society*, 1871, p. 528.

bottom, whether in 2000 or 20 fathoms." The float which Captain Spratt, after experience, found to answer best, was one of thin copper or block-tin, suspended like a kite, his observations being in this respect the same as those at the Cromarty Firth, while the boat which was moored at the Cromarty Firth, in order to obtain the relative speeds of the surface and under current floats, fulfilled the object of his buoy moored with a sinker.

Under-current
floats used by
Mr. Mitchell.

Mr. Henry Mitchell, of the United States Coast Survey, describes an instrument used by him for that purpose. It consisted of a tin cylinder, a few inches in diameter, and long enough to reach from the surface nearly to the bottom. Tubes 40 feet in length were used for this purpose. They were 3 inches in diameter, made in separate sections, air-tight, but with stop-cocks for letting in water, that they might be practically filled, so as to sink to the proper depth. As the tube drifted nearly upright in the water, with its top protruding a few inches above the surface, its velocity indicated the *mean motion* of the stream. If it leant backwards or forwards, it showed that its foot rested on a stratum that had greater or less motion than the surface drift; and if its angle of direction differed from that of the surface log, the action of an under-current was recognised, whose course was at variance with that of the surface drift.

Mr. Mitchell also says, that very good " results have been obtained by using two hollow copper globes of 2 feet diameter each, connected by $\frac{1}{8}$ inch wire rope. The sinking globe is filled with water, but the other is loaded only enough to sink nearly to its pole. The upper globe has a long line secured to it, and its motion is recorded at the same time that an observation is made with the surface log, like the compound float used by Dr. Anderson at the Tay, as described at page 99.

Mr. Mitchell says, "Let us suppose that the two globes present equal effective areas (great circles) to the drifts in which they swim, then their velocity will be a true mean of the rates of the surface and under currents; *i.e.* $\frac{1}{2}(x+y)$ where x and y represent respectively these rates. The velocity of the under-current may therefore be found by subtracting the surface rate from twice that of the connected globes."

This formula no doubt gives the mean for the velocities of the two strata in which the balls are floating, and it would give the mean for the whole column of water, provided there is a regular gradation between these two observed velocities, but it does not provide for any inequality of velocity, or for any anomalous velocity, such as has been stated to exist at the Dee and the Cromarty Firth. This objection might perhaps to some extent be removed if it were practicable to suspend balls similar to those used by Mr. Mitchell, at short intervals on the wire rope. But for engineering purposes, the object of ascertaining the under-currents has, in my experience, always been to calculate the discharge; and it is obvious that for this purpose we must determine the thickness of the different strata moving at different velocities, so as to ascertain the different sectional areas to which the velocities apply, and this not at *one* but at *several* points on the cross section of the channel or passage through

which the current was flowing. Until we have some
method of ascertaining the velocities at different depths,
and the sectional areas corresponding to these velocities,
it is not possible to arrive at the discharge, and all observation on the strength and duration of under-currents
must be regarded by the engineer, in making calculations,
to be simply approximate.

The remarkable under-currents of the Cromarty Firth
are mainly, if not altogether, due, I believe, to the configuration of the bottom, and the circumstances under
which the tidal wave approaches and recedes from the
shore. A powerful oceanic under-current during flood-
tide in a stratum of water of high specific gravity and low
temperature, setting dead along the coast, would naturally creep along the rising bottom of the sea, and flow
into the deep inlet of the firth, mingling imperfectly with
the surrounding water, maintaining its character of a distinct stream, and increasing the under-velocity of the
flood-tide; and if we suppose a similar rapid counter-
current to sweep along the coast at ebb-tide, its tendency would be to draw off the lower stratum of denser
and colder water, and thus to increase the velocity at
or near the bottom during ebb-tides. Of the existence
of such *distinct* ocean currents, some at great depths,
and others superficial, maintaining their character, and
mingling slowly with the surrounding ocean, there are
many striking examples; among others, the surface-
current of the Gulf Stream, which, flowing from the
Gulf of Mexico, skirts the coast of the United States,
and can be traced as a distinct body of water by its

difference of temperature as far as the banks of New-foundland. In the month of July I found the temperature of the sea, as tested at various points between the shore of America and the edge of the Gulf Stream, to average 60° Fah.; while in lat. 41° N., long. 61° 52′ W., the vessel being in the track of the Gulf Stream, the temperature of the water was 70°. After leaving the influence of the Gulf Stream, the temperature within a few hours' sail fell to 60°, which was the average of the observations made during the remainder of the voyage to the English Channel, ascertained as accurately as the facilities granted to a passenger by a packet-ship permitted.

The cause of ocean currents is obscure. They no Cause of under-currents doubt are occasionally caused or increased by gales of wind. But, as I pointed out in the first edition of this book, no current can be generated without a difference of *head*, which again may be produced either by a difference of level in water of the same density, or by a difference of specific gravity in columns of water of the same height. The examples I have given of the differences of levels existing in rivers and estuaries at certain states of the tide, afford sufficient proof of the existence of currents from that cause. It is not unusual to employ the expression *indraught*, to describe the flow of water into a bay or creek, and to hear it used so as almost to imply the existence of some inherent attraction in the bay or creek for the water which flows into it. But the flow of water in all such cases is caused by the pressure due to difference of level or density, or both combined; and when the bay or creek gets filled up, and its surface attains a sufficient height to balance the pressure of the source of its supply, or the momentum of the moving column of water where converging shores cause the level of the water to rise, the indraught disappears.

DENSITY OF SALT AND FRESH WATER.

Before leaving this part of the subject, however, I must say something more as to the separation of the *under* from the *superficial* strata of water, and explain that other species of disturbance which is due to the different density of salt and fresh water.

Methods for obtaining specimens of water from different depths. The first observations on this subject to which I shall refer were those made by my father on the river Dee, in Aberdeenshire, in the summer of the year 1812, when engaged in surveying that river in reference to a salmon-fishing case.[1] "He observed in the course of his survey that the current of the river continued to flow towards the sea with as much apparent velocity during flood as during ebb tide, while the surface of the river rose and fell in a regular manner with the waters of the ocean.   He was led from these observations to inquire more particularly into this phenomenon, and he accordingly had an apparatus prepared, under his directions, at Aberdeen, which, in the most satisfactory manner, showed the existence of two distinct layers or strata of water ; the lower stratum consisting of salt or sea water, and the

[1] Report to the Earl of Aberdeen and the other Proprietors of the " Raik " and " Stell " Fishings of the River Dee, at Aberdeen, by Robert Stevenson, Civil Engineer, Edinburgh, Feb. 1813 ; and Stevenson's *Bell-Rock Lighthouse*, p. 79.

upper one of the fresh water of the river, which, from its
specific gravity being less, floated on the top during the
whole of flood as well as ebb tide.   The apparatus consisted of a bottle or glass jar, the mouth of which measured about $2\frac{1}{2}$ inches in diameter, and was carefully
stopped with a wooden plug, and luted with wax ; a hole,
about half an inch in diameter, was then bored in the
plug, and to this an iron peg was fitted.   To prevent
accident in the event of the jar touching the bottom, it
was coated with flannel.   The jar so prepared was fixed
to a spar of timber, which was graduated to feet and
inches, for the conveniency of readily ascertaining the
depths to which the instrument was plunged, and from
which the water was brought up.   A small cord was attached to the iron pin for the purpose of drawing it at
pleasure for the admission of the water.   When an experiment was made, the bottle was plunged into the water ;
by drawing the cord at any depth within the range of the
rod to which it was attached, the iron peg was lifted or
drawn, and the bottle was by this means filled with
water.   The peg was again dropped into its place, and
the apparatus raised to the surface, containing a specimen
of water, of the quality at the depth to which it was
plunged.   In this manner the reporter ascertained that
the salt, or tidal water of the ocean, flowed up the channel of the river Dee, and also up Footdee and Torryburn,
in a distinct stratum next the bottom and under the
fresh water of the river, which, owing to the specific
gravity being less, floated upon it, continuing perfectly
fresh, and flowing in its usual course towards the sea, the

only change discoverable being in its level, which was raised by the salt water forcing its way under it. The tidal water so forced up continued salt, and when the specific gravities of specimens from the bottom, obtained in the manner described, were tried, and compared with those taken at the surface, by means of the common hydrometer of the brewer (the only instrument to which the reporter had access at the time), the lower stratum, when compared with that at the surface, was always found to possess the greater degree of specific gravity due to salt over fresh water."

The instruments now used for obtaining water from different depths are more perfect in their construction than the original instrument used for that purpose at the Dee, which was made for a temporary purpose. Various constructions have more recently been tried for experimenting on this subject, by Scoresby, Sabine, Dr. Marcet, and others, but for ordinary engineering inquiries I can confidently recommend the instruments I am about to describe, which I have termed hydrophores, and have extensively employed in engineering surveys.

Hydrophores, and manner of using them.

Fig. 17 represents a hydrophore used for procuring specimens of water from moderate depths, drawn on a scale of one-tenth of the full size. It consists of a tight tin cylinder, letter *a*, having a conical valve in its top, *b*, which is represented in the diagram as being raised for the admission of water. The valve is fixed *dead*, or immoveable, on a rod working in guides, the one resting between two uprights of brass above the cylinder, and the other in its interior, as shown in faintly dotted lines.

The valve-rod is by this means caused to move in a truly vertical line, and the valve attached to it consequently fills or closes the hole in the top of the cylinder with greater accuracy than if its motion was undirected. A graduated pole or rod of iron, c, which in the diagram is shown broken off, is attached to the instrument, its end being inserted into the small tin cylinder at the side of the large valve or water cylinder, and there fixed by the clamp screws shown in the diagram; the bottom of the water cylinder may be loaded with lead to any extent required, for the purpose of causing the apparatus to sink; but this,

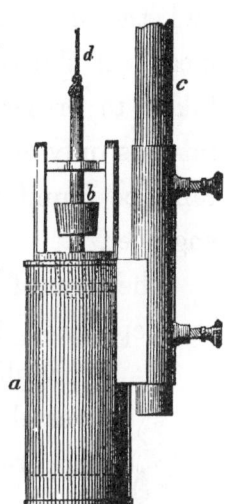

FIG. 17.

when an iron rod is used for lowering it, is hardly necessary. The spindle carrying the valve has an eye in its upper extremity, to which a cord is attached for the purpose of opening the valve when the water is to be admitted, and on releasing the cord, it again closes by its own weight. When the hydrophore is to be used, it is lowered to the required depth by the pole which is fixed to its side, or if the depth be greater than the range of the pole, it is loaded with weights and let down by means of a rope so attached as to keep it in a vertical position. Care must be taken, while lowering or raising it, that the small cord by which the valve is opened be allowed to hang perfectly free and slack. When the apparatus has been lowered as far as is required, the small cord is pulled, and the vessel is immediately filled with the water

which is to be found at that depth. The cord being then
thrown slack, the valve descends and closes the opening,
and the instrument is slowly raised to the surface by
means of the rod or rope, as the case may be, care being
taken to preserve it in a vertical position. This apparatus is only applicable to limited depths, but will generally be found to answer all the purposes of the civil
engineer.

The form of hydrophore represented in fig. 18 is used
in deep water, to which the small one just described is

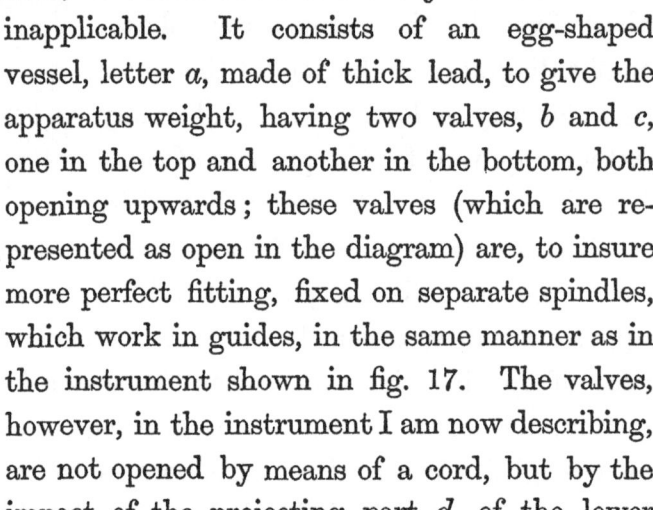

inapplicable. It consists of an egg-shaped
vessel, letter a, made of thick lead, to give the
apparatus weight, having two valves, b and c,
one in the top and another in the bottom, both
opening upwards; these valves (which are represented as open in the diagram) are, to insure
more perfect fitting, fixed on separate spindles,
which work in guides, in the same manner as in
the instrument shown in fig. 17. The valves,
however, in the instrument I am now describing,
are not opened by means of a cord, but by the

Fig. 18.   impact of the projecting part d, of the lower
spindle on the bottom, when the hydrophore is sunk to
that depth. By this means the lower valve is forced
upwards, and the upper spindle (the lower extremity of
which is made nearly to touch the upper extremity of the
lower one, when the valves are shut) is at the same time
forced up, carrying along with it the upper valve which
allows the air to escape, and the water rushing in fills the
vessel. On raising the instrument from the bottom both

valves again shut by their own weight and that of the
mass of lead d, which forms part of the lower spindle.
The mode of using this hydrophore is sufficiently obvious;
it is lowered by means of a rope, made fast to a ring at
the top, as is shown in fig. 18, until it strikes on the
bottom, when the valves are opened in the manner
described, and the vessel is filled; on raising it the valves
close, and the vessel can be drawn to the surface without
its contents being mixed with the superincumbent water
through which it has to pass. This instrument weighs
about half a hundredweight, and has been easily used in
from 30 to 40 fathoms water in making engineering
surveys, and could no doubt, if necessary, be employed
for much greater depths. It is represented in the cut on
a scale of one-twentieth of the full size.

The hydrophore employed by the Scientific Explora- Hydrophore
used in deep-sea
tion of the Deep Sea, in 1870, was suggested by the explor tions.
hydrographer to the Admiralty, and consisted of a strong
cylinder of brass, 26 inches long and 2·3 inches diameter,
holding about 60 oz. of water in the disk which closes it;
at each end there is a circular aperture, into which a
conical valve is accurately fitted. While this bottle is
descending through the water with the sounding apparatus the valves readily yield to the upward pressure, and
a continuous current streams through it, but so soon as
the descent is checked, either by the arrival of the apparatus at the bottom, or by a stop put on the lines from
above, the valves fall into their places, and there enclose
the water that may fill the bottle at the moment. The
working of this simple apparatus was found to be entirely

I

satisfactory. It is obvious however that the deep-water hydrophore would not be found so useful for the smaller depths requisite in engineering surveys as the two forms of the instrument which I have described and used since 1842.[1]

Dr. Marcet, in a paper on the specific gravity of sea-water, in the *Philosophical Transactions* for 1819, describes an instrument he used for obtaining water from the bottom, consisting of a cylindrical vessel with an aperture at either end fitted with valves opening upwards. These valves when closed were secured by springs, but were so made as to be kept open by a weight acting on the springs, and suspended below the vessel. When this weight touched the bottom the springs were relieved, and closed the valves, which remained shut, and enclosed the specimen of water. I do not think however that this arrangement is so simple as that shown in the preceding figure, which, as used in moderate depths, I never found to fail.

In all these experiments, the water being emptied into bottles, is corked up, sealed, and labelled with certain numbers, which should be entered in a book containing remarks as to the place of observation, time of tide, and such other particulars as, from the nature of the inquiry, seem to deserve notice, and the water thus preserved may be subjected to analysis.

The appearance of fresh or brackish water floating on the surface of the sea, as described at the Dee at Aberdeen, is no doubt familiar to most observers. It occurs

---

[1] Preliminary Report of the Scientific Exploration of the Deep Sea, by Dr. Carpenter, J. Gwyn Jeffreys, F.R.S., and Wyville Thomson, LL.D., 1870.

indeed more or less at the mouths of all rivers, being most apparent when they are in flood, from the browner tinge given to the water, which is sometimes discoloured many miles at sea. Father Manuel Rodriguez, a Spanish Jesuit, speaking of the Amazon, says[1]—" This river is like a tree ; its roots enter far into the sea, as into the land. It communicates to it a flavour, so that at 80 leagues within the sea its waters are seen, and taste sweet, and in a semicircle of 100 leagues in circumference they form a gulf not the least degree brackish, so that sailors call it the fresh sea,"—a statement which might almost seem incredible, had not Sir Edward Sabine, in 1827, found something which goes far to bear out the correctness of the Spaniard's account, and which he describes in the following words :—"At 10 A.M. on the 10th of September, whilst proceeding in the full strength of the current, exceeding, as already noticed, 4 knots an hour, a sudden and very great discoloration in the surface-water ahead was reported from the mast-head, and from the very rapid progress which the ship was making was almost immediately afterwards visible from the deck. Her position in 5° 08′ north latitude, and 50° 28′ west longitude, sufficiently apprised us that the discoloured water which we were approaching could be no other than the stream of the river Amazon, preserving its original impulse at a distance of not less than 300 miles from the mouth of the river, and its waters being not yet wholly mingled with those of the ocean of greater specific gravity, over the surface of which it had pursued its course.

[1] *El Maranam y Amazonas.* Madrid, 1684, p. 18.

"We had just time to secure some of the blue water of the ocean for subsequent examination and to ascertain its temperature before we crossed the line of its separation from the river-water, the division being as distinctly preserved as if they had been different fluids.

"The direction of the line of separation was N.W. by N., rather northerly; great numbers of gelatinous marine animals, species of the genus *Physalia*, were floating on the edge of the river-water, and many birds were fishing apparently on both sides of the boundary."[1]

I have occasionally seen these brownish-coloured patches at a considerable distance from the coast, and on one occasion, in the Pentland Firth, on drawing a bucket of this brownish water, and comparing it with that of the sea after passing through the patch, it was found to be distinctly *brackish*. It is well known to the crews of "welled" smacks employed in cod-fishing on our coasts that they invariably lose a portion of their live stock if they happen to encounter what they term a "fresh," which is believed by them to be a brackish portion of the sea, caused, no doubt, by the imperfect mixture of the fresh water discharged from rivers.

Specific gravities of salt and fresh water.

In subjecting waters for examination to ascertain the proportion of fresh and sea water, two methods may be adopted—first, by taking their specific gravity; or secondly, by evaporating a certain quantity, and ascertaining the amount of saline matter left. The result may, in either case, be to some extent affected by the quantity of vegetable matter in suspension, but it is sufficiently

[1] *Philosophical Magazine*, vol. lxvii.

accurate for most engineering inquiries. The specific gravity test is the most convenient, and that which is generally adopted, the specific gravity being taken with an ordinary hydrometer; the standard employed is distilled water when at the temperature of 62° Fahrenheit as 1000. Dr. Marcet says that in general the waters of the ocean, whether taken from the bottom or the surface, contain most salt in places where the sea is deepest and most removed from land, and in his tables of experiments he gives the specific gravity of some ocean samples as high as 1030·9. From various observations, made at different parts of our coasts, I am disposed to state the specific gravity in these localities at 1026. I have found that in any experiments made in our rivers and estuaries the densities varied from 1000 to 1026, and occasionally during flood-tide the difference in specific gravity between the surface and bottom is very striking, as pointed out in Mr. Stevenson's experiments of 1812 at Aberdeen.

Sizes of rivers proportional to the extent of country drained—The Mississippi an
example of a large river—Description of its navigation, currents, and discharge—Works proposed for its improvement—Means used for rendering the
upper portions of small rivers navigable, by stanches, dams, and locks—Improvements of upper portions of Continental rivers, such as the Rhine and
Danube.

In following out the division of the subject proposed
at the conclusion of Chapter III., I have, in treating of
rivers, to consider in the first place what has been termed
the upper or " river proper" compartment. And as regards the use made of such streams for the purposes of
navigation, or the works calling for the engineer's assist-
ance to render them navigable, there is not much to refer

Sizes of rivers proportional to the extent of country drained.

to in this country. The magnitude of a river and its use-
fulness for navigation may be said, under certain conditions, to be proportional to the extent of country which is
drained. Thus in continents we find rivers of great
magnitude, fed by the drainage of vast tracts of surrounding land, rolling their contents in a broad, deep current
to the ocean, and affording a highway for vessels of the
largest class to pursue their course for hundreds of miles
into the interior of the country. Of such is the Mississippi,
which maintains, for a distance of nearly 1200 miles
above New Orleans, an average breadth of 3300 feet, and
a depth of 115 feet. The Ohio, which joins it at this

place, is navigable to Pittsburgh, where I have seen from thirty to forty large-sized steamers lying at the quays of that truly inland port, which were all engaged in trading to New Orleans, on the Gulf of Mexico,[1] being a river-navigation of upwards of 2000 miles.

In considering the improvement or maintenance of such a navigation as this, the engineer has to deal chiefly with the control of the enormous body of water which it discharges. His difficulty does not consist in deficient depth or breadth of navigable channel, but in the magnitude of the floods with which he has to contend, and the provision he has to make for retaining them within such limits as to secure the safety of the surrounding district.

In less extended tracts of country than the valley of the Mississippi, the rivers are proportionally smaller; and when we come to consider our own island, we find that its area and drainage are only sufficient to supply streams of the smaller class.

THE MISSISSIPPI.

The Mississippi is the most gigantic river-navigation in the world, and some facts as to its navigation, taken from the elaborate report of Mr. Charles Ellet,[2] which was made to the Government of the United States, cannot fail to interest the engineer, and will not, I am sure, be considered out of place in this work; for, although they cannot be said to apply to British, or even Continental rivers, they will at least best serve to show by comparison the smallness of our own rivers when I come to speak of them.

---

[1] *Sketch of Civil Engineering of North America*, by David Stevenson, C.E.

[2] *The Mississippi and Ohio Rivers*, by Charles Ellet, Philadelphia, 1853.

It appears, from the information given in Mr. Ellet's work, that the Mississippi varies from 2200 to 5000 feet in width, the average width being assumed as 3300 feet. It is from 70 to 180 feet in depth, the average being 115 feet. The area of the cross section varies from 105,544 square feet to 268,646 square feet, the average being 200,000 square feet. The length, from its junction with the Ohio to the Gulf of Mexico, is 1178 miles, and its average fall at full water is $3\frac{1}{4}$ inches per mile, and in absence of floods (or during summer and autumn) $2\frac{8}{10}$ inches per mile. The length of the Ohio, from its junction with the Mississippi to Pittsburgh (the head of the navigation for large vessels), is 975 miles, and the average inclination is about $5\frac{1}{5}$ inches per mile. From Pittsburgh to Olean Point, the head of the navigation for small vessels, the distance is 250 miles, and the inclination 2 feet 10 inches per mile. When the water is high, even steamboats have ascended to Olean Point, which is 2400 miles from the Gulf of Mexico; and in doing so, have had to overcome a current which at some places runs with a velocity of 5 miles per hour. Generally speaking, vessels have no difficulty, in the lower or more open part of the stream, in avoiding the strength of the currents by keeping in-shore. But in the Ohio much inconvenience is felt during dry seasons from the currents at certain parts of the river; and I have seen a steamer unable to overcome them until assisted by a warp attached to an anchor dropped ahead of the vessel, in the middle of the channel, by which, after considerable detention, she was "warped through the rapid;" there are no such shoals, however, in the Mississippi, nor indeed below Louisville on the Ohio.   The discharge of the Mississippi is computed by Mr. Ellet, at high water, at 1,280,000 cubic feet per second ; and its drainage he estimates at 1,226,600 square miles.   When the autumnal rains set in, the river rises above its summer level to the enormous extent of about 40 feet at the mouth of the Ohio, and 20 feet at New Orleans.   In investigating the physical characteristics of this mighty stream, Mr. Ellet found—1st, That the average surface velocity in the centre of the river was 5 miles per hour, and occasionally the speed reached 7 miles per hour ; 2d, By using under-current floats, he found that the speed of a float, supporting a line of 50 feet long, was always greater than that of the surface float—the average increase of velocity being 2 per cent. ; 3d, The results of the experiments made lead him to conclude that the mean velocity of the Mississippi is about 2 per cent. greater than the mean surface velocity ; 4th, In coming to this conclusion, no account is taken of such observations as show remarkable under-currents, the velocity of which were in some places found to be 17 per cent. and $20\frac{1}{3}$ per cent. greater than the surface velocities ; 5th, While the mass of water which the channel of the Mississippi bears is running downwards with a central velocity, the current next the shore is sometimes found to be running upwards, or in the opposite direction, at the rate of 1 to 2 miles per hour ; 6th, While the water is running downwards in the one side of the river, it is often found with an appreciable slope, and visible current running upwards on the other side of the river ; 7th, The

surface of the river is therefore not a plane, but a peculiarly complicated warped surface, varying from point to
point, and inclining alternately from side to side. After
considering all the conflicting results derived from his
investigations, Mr. Ellet, in order to obtain the mean
velocity and discharge of the river, employed the formula
as already noticed :—

$$V = \frac{8}{10}\sqrt{df} + \frac{df}{20}$$
$$M = 0 \cdot 8\ V$$
$$Ma = D$$

where V = the velocity of central surface current in feet per
          second,

    d = maximum depth of river in feet at place of observation,

    f = slope of surface in feet per mile,

    M = the mean velocity in feet per second,

    a = area of cross section of river in feet,

    D = discharge of river in cubic feet per second.

In discussing the various formulæ for velocities and
discharges, we have already seen, at page 109, that the
formula applied to the Mississippi by Mr. Ellet does not
apply to such rivers as the Tay, or to smaller water-
courses ; and, indeed, until the result which he has given
has been compared with the discharge obtained by actual
measurement of the velocities at different parts of the
cross section, we do not think that the discharge of the
Mississippi, which has been calculated by Mr. Ellet, can
be relied on as accurate.

The "peculiarly warped" form assumed by the surface from side to side and from point to point of the
water, is an interesting feature, which renders any

gauging of the discharge exceedingly difficult, and may, I think, be accounted for on the "principle of the conservation of forces," which is more fully noticed in the chapter on tidal rivers. If we can imagine an upright wall opposed as a barrier across the column of water moving down the Mississippi, so as to arrest its progress, the surface of the water would not rise equally against the face of the wall or barrier from side to side of the river. The rise or elevation of the surface would be highest when the momentum of the water was greatest. Thus, for example, in a bridge the water rises highest on the cut-waters of those piers which stand in the greatest depth and strongest current, and so at the Mississippi the column of water, though not stopped by a solid obstruction, is nevertheless opposed by numerous contractions, abrupt bends, and islands causing the surface to rise *unequally*, and thus to generate counter or side currents and all the disturbance due to unequal pressure which Mr. Ellet describes.

The chief object of the investigations made by Mr. Ellet was the prevention of floods, which have recently increased both in number and extent. This he attributes—

First, To extended cultivation, by which evaporation is supposed to be diminished, the drainage increased, and the floods hurried forward more rapidly into the country below.

Second, To the extension of the embankments along the banks of the Mississippi and its tributaries, by which water that was formerly allowed to spread is now confined to the channel of the river.

Third, To what are termed cut-offs, or straight cuts, by which the distance is shortened, and the slope and velocity increased, so that the water is brought down more rapidly from the country above.

Fourth, To the gradual extension of the delta into the sea, so as to lengthen the lower course of the river, to diminish the slope and velocity, and thus to throw back the water on the land above.

The works suggested for protecting the country against floods are—

First, More sufficient embankments.

Second, The prevention of further cut-offs, or works for straightening the upper parts of the tributaries of the river.

Third, The enlargement of the seaward channels or outlets. And

Fourth, The creation of large artificial reservoirs, by placing dams across the outlets of the lakes or distant tributaries, so as to compensate for the loss of the natural overflow of the water, which is checked by the embankments for protecting the country in the lower part of the river.

I am not aware whether any of Mr. Ellet's suggestions have been carried out.

His report had reference chiefly to the question of drainage,—an important one in a district where the flood-waters of the river attain an elevation considerably higher than the adjoining country. Mr. Ellet says that the river carries at all times a vast amount of earthy matter, which the current is able to carry forward as long as the river is

confined to its channel; but when the water overflows its
confining embankments, it deposits the particles in suspension on either side, leaving the heavier matter nearest
to the river; consequently the borders of the river, which
receive the first and heaviest deposit, are raised by successive floods above the general level of the delta, and
ultimately assume a cross section similar to that shown
in fig. 19, where the horizontal dotted line shows that the
surface of the river is higher than the level of the land
on either side. Mr. Ellet gives this as an average section,
obtained from a number of surveys made at the lower part

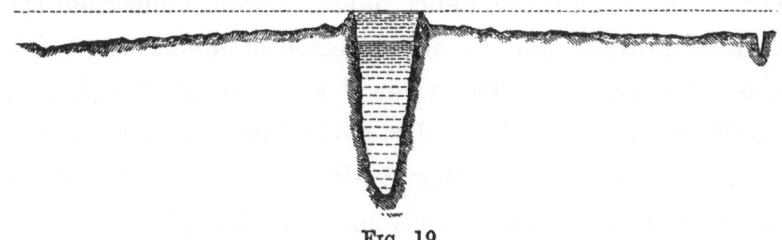

Fig. 19.

of the delta, and states that the land is from 18 to 20 feet
lower than the river.

The Mississippi and its tributaries drain the whole of
the North American continent, which extends from north
to south between the Great Northern Lakes and the
Gulf of Mexico, and from east to west between the ranges
of the Alleghany and Rock Mountains. These fertile
valleys include nine of the United States of America.
The geological formation of the country shuts up this
immense tract of land from any direct communication
with the seas which wash the eastern and western coasts
of the continent; for if we trace upwards in their courses
of many hundred miles through the eastern States, these

numerous large navigable rivers which discharge into the Atlantic, we find them holding the character of streamlets, long before we penetrate even to the range of these fertile valleys; and on the western coast of the country the range of the Rocky Mountains, extending along the shores of the Pacific, presents an insurmountable barrier to any direct water-communication with that ocean. The Mississippi, however, and its numerous tributaries, afford a perfect and easy access to the remotest corner of these regions. The source of the river is said to have been discovered, in the year 1832, to the westward of the Great Lakes, at the distance of about 3000 miles from the Gulf of Mexico, and at an elevation of about 1500 feet above its surface. The river flows from its source as a small stream, and gradually gathering strength, passes over the falls of St. Anthony, after which at every stage of its course it gains accessions of strength from the numerous small rivers that pour in their tributary streams from all directions, until it is joined by the great Missouri. The character of its water, formerly clear and tranquil, is here completely changed, and the combined streams of two rivers flow on in a deep and muddy current. The Ohio, the Arkansas, the Red River, and many other large streams, fall into this giant of rivers, which, swelled by the waters of its various tributaries, at last pours into the Gulf of Mexico.[1] The aggregate length of the various tributaries of the Mississippi has been computed to be upwards of 44,000 miles.

But leaving the class of giant rivers, of which the

[1] Stevenson's *American Engineering.*

Mississippi is a type, and which, without the help of artificial works, afford ample width and depth for extended lines of navigation, we shall consider the means used for rendering the upper portions of the smaller class of rivers navigable.

Telford says,[1] " A river, in its natural current, is more or less deep from circumstances which need not here be described, and its navigation is usually impeded by shallows and rapids—inconveniencies which the ingenuity of man has striven to overcome, ever since his boats became too large and too heavy for portage, as is still in use for conveyance by canoes in the North American fur-trade. The first expedient which occurred was to thrust the boat as nearly as possible to the rapid, and having well fastened her there, to await an increase of water by rain; and this was sometimes assisted by a collection of boats, which, by forming a kind of floating dam, deepened the water immediately above, and threw part of the rapid behind themselves. This simple expedient was still in practice at Sunbury, on the river Thames, since the beginning of the present century; and elsewhere the custom of building bridges almost always at fords, to accommodate ancient roads of access, as well as to avoid the difficulty of founding piers in deep water, afforded opportunity for improvement in navigating the rapid formed by the shallow water or ford; for a stone bridge may be formed into a lock or stoppage of the river by means of transverse timbers from pier to pier, sustaining a series of boards called paddles, opposed to the strength of the

[1] Telford's *Life*, p. 57.

current, as was heretofore seen on the same river Thames, where it passes the city of Oxford at Friar Bacon's Bridge, on the road to Abingdon. Such paddles are there in use to deepen the irregular river channels above that bridge ; and the boat, or collected boats, of very considerable tonnage, thus find passage upwards or downwards, a single arch being occasionally cleared of its paddles to afford free passage through the bridge."

Stanches. Sir William Cubitt also says, there were thirteen old stanches, as they were called, on the Stour, in Essex. These consisted of two substantial posts, which were fixed in the bed of the river at a sufficient distance apart to permit a boat to pass easily between them, and connected at the bottom by a cross cill. Upon one of these posts was a beam turning on a hinge or joint, and long enough to span the opening. When the "stanch" was used, the boatman turned the beam (which was above the level of the water) across the opening, and placed vertically in the stream, a number of narrow planks resting against the bottom cill and the swinging beam, thus forming a weir which raised the water in the stream about 5 feet high. The boards were then rapidly withdrawn, the swinging beam was turned back, and all the boats which had been collected above were carried by the flow of water over the shallow below. By repeating this operation at given intervals, the boats were enabled to proceed a distance of about 23 miles in two or three days.

Still-water navigation. This primitive system, which was at one period very common in England, has been superseded by throwing permanent dams across the river, so as to convert its

channel into a series of deep-water reaches, and the boats
pass from one reach to the other by means of side-cuts
with locks.   This plan, which in America is called " still-
water navigation," has been extensively carried out on the
rivers of that country.   I saw a good specimen of it on
the Schuylkill, in Pennsylvania.   That river was rendered
navigable by thirty-four dams constructed in the bed of
the stream, so as to raise the level of the water and convert the river into thirty-four reaches of navigable water,
varying in length according to the rise in the river's bed.
The barges pass from the different reaches through a short
side-cut, in which there is a canal lock of the ordinary
construction.   The navigation is upwards of 100 miles
in length, and was navigated by boats of about 60 tons
burden.   The same plan has been carried out on a pretty
large scale by the late Mr. Rendel and Mr. Beardmore,
for the improvement of the river Lee, and by Sir William
Cubitt on the upper part of the Severn, where the river
has been divided into four reaches, having a depth of 6
feet, with side-cuts and locks having a lift of 8 feet each.

It must be obvious that the works for forming slack-
water navigation closely resemble the ordinary canal works
which have been already described, and the side cuts and
locks for passing vessels from reach to reach of the river
are identical, and require no further notice.

But the operation of damming up the river is impor- Weirs or Dams.
tant, and cannot be passed over without special notice.
A river dam is a work in all cases demanding careful consideration, not only as regards its safe construction so as
to resist the force of the stream, but also with reference

K

to its effect in opposing the free discharge of the water, and causing the land above it to be flooded during heavy rains.

The dams on the Schuylkill navigation, and, indeed on many of the American rivers, are formed of timber framework filled with rubble. That on the Schuylkill at Philadelphia, which served both for the navigation and for

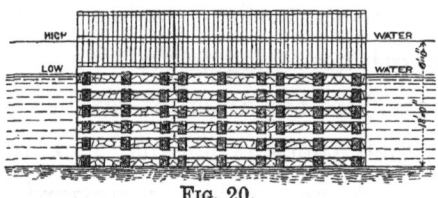

FIG. 20.

supplying water to drive the wheels of the Philadelphia Water-works, was formed in separate compartments or frames, each of which was 20 feet in breadth. These, after being framed together, were filled with stones, and sunk in the line of the dam. Fig. 20 is an elevation, and

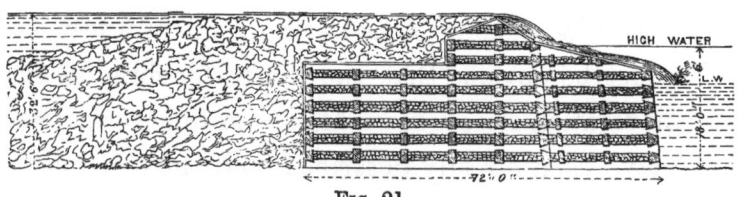

FIG. 21.

fig. 21 is a section of the dam, from which its construction will be easily understood. The cribs were formed of logs of wood measuring 18 to 20 inches, connected together by strong dovetailing. The size of the framework in the direction of the stream was 72 feet. The planking on the top was 6 inches thick. The upper parts of the cribs

were connected together so as to form one continuous
structure, and the whole was backed by a large mass of
heavy rubble.   This dam withstood a flood, when the
river Schuylkill was passing over it in a solid body of
water, 7 feet 11 inches in depth without sustaining any
injury.

The dams constructed by Sir W. Cubitt on the
Severn are shown in fig. 22.   They are formed of pile

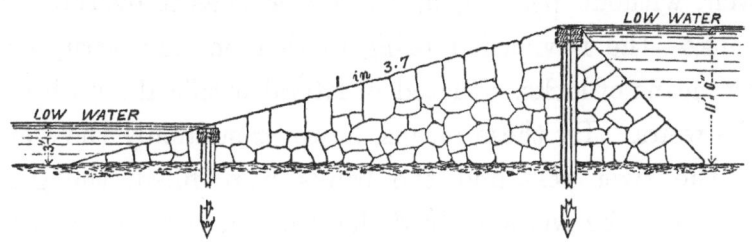

FIG. 22.

work and rough masonry, and have also withstood the
floods.

But the important question of flooding still remains
to be noticed, and as Sir W. Cubitt very carefully studied
that subject, I cannot do better than give what he laid
before the Institution of Civil Engineers[1] as the result of
his experience :—

"The problem proposed (he said) was this : a river
which, from its source and the country it passed through,
was liable to sudden changes of character—at one time
running deep and rapid like a mountain stream, and then
suddenly subsiding, and becoming so shallow as to impede
the navigation, was required to be so improved as to
retain the waters, that the minimum depth, in times of

[1] Vol. v. p. 348.

drought, should be sufficient for the traffic, and yet that the passage for the flood-waters should be so unimpeded, that they should pass off, without unduly flooding the lands on either side of the stream, or injuring the drainage. This was done by placing certain weirs obliquely across the main stream, their length being such as to permit of the free passage of a body of water equal to the entire transverse sectional area of the river above the weir, without penning up the water so as to overflow the banks—the navigation being carried on uninterruptedly by means of locks, situated in lateral artificial cuts beside the weirs. By these means, however suddenly the water in the river rose and the banks were filled, the great length of the weirs enabled the flood to pass away, without reducing the average depth in the channel, yet allowing for the free drainage of the land above.

" In placing a weir directly square across a river, a considerable portion of its section must be blocked up, and the water would be penned back, in proportion to the actual height of the weir and the area of the channel. The water could never flow over that weir in a sheet wider than the channel, consequently, the depth upon the weir must be greater, and the tendency must be to block up the water, and even with a river *bank-full* an obstruction must exist. It would be seen

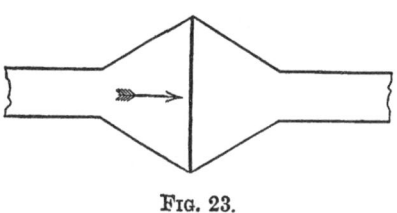

Fig. 23.

by fig. 23, that if the breadth of the channel were extended to three times its width, above and below, and a transverse weir placed, the same quantity of water would pass over, in a thinner sheet, leaving a comparatively tranquil pool above the weir, which, if extended to a lake, would present the same appearance as the Lake of Geneva —comparatively still water, with a rapid river above and below it.   It must be evident that the weir in this case offered no obstruction, as the water was enabled to pass more freely than along the river, either above or below.

" Fig. 24 showed that the same end might be attained without the expense of cutting away the land, by placing a weir of the

FIG. 24.

same length in an oblique direction across the stream, without unduly widening the channel.   The same quantity of water would pass in times of flood, and the velocity of the stream would be maintained more equably, than by any direct transverse obstruction.   The greater the obliquity the better would be the effect ; but, as a general rule, three times the direct width of the channel would, he thought, be an ample proportion,—in fact, it would seldom be necessary to give more than double the direct width.   As a simple rule, it might be stated, that when the rectangle formed by the length of the weir and its depth below the flood limit, equalled the rectangle of the river up-stream of the weir, within the same flood limits, then similar floods would not rise higher above these limits than before the weirs were placed.

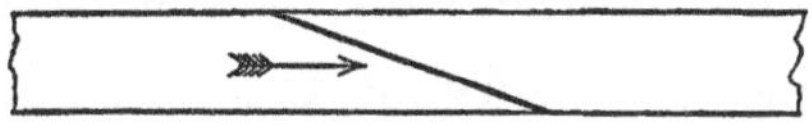

" It had been attempted to be shown that the velocity of the under-current would be checked, and that a deposit would take place ; in fact, that the effect would

be analogous to filling up the bed of the river to the
height of the weirs, and that the descending column, for
a considerable distance up the river, would be materially
checked. The contrary, however, had been the result.
The surface of the water had been prevented from rising
during floods; the bed of the river had not been raised,
although dredging had not been resorted to; and the
drainage of the upper country had been fully maintained.
The under-current received no check, as, from the obliquity of the weir, the stream was merely turned aside,
and the body of the water rose gradually at its initial
velocity.

"Weirs, of the shape of a horse-shoe, and with an
acute angle pointing up the stream, had also been tried;
but there were practical objections to both these forms,
as not leaving a free space within them for the overfall
of the water, which there formed eddies, frequently scouring out the bed of the channel at the foot of the weir."

Much difference of opinion has been expressed by
engineers on the question raised by Sir W. Cubitt as to
the relative obstruction to a river's flow presented by
transverse and oblique weirs, and I am not aware that
we possess any data determined by actual experiment.
I think all engineers, however, agree on the advantage of the oblique weir in facilitating the discharge;
and meantime it is satisfactory to know that the rule
adopted by Sir William Cubitt, of making the weir of *such
length as that the rectangle formed by its length and its
depth below the flood line shall be equal to the rectangle
of the river above the weir within the same flood limits,*

has given most satisfactory results on the Severn, where weirs constructed on that principle have not increased the flooding of the river's banks.

Oblique weirs have also been employed with great advantage in the Shannon. Mr. Rhodes states that the weirs at Killaloe and Meelick are each 1100 feet in length, that in summer water there is a flow of 6 inches, and in high floods from 2 feet 6 inches to 2 feet 8 inches over their crests, and they do not produce flooding of the lands above.

RHINE AND DANUBE.

The works executed for the improvement of the upper portions of many of the continental rivers, such as the Rhine and the Danube, are very different in their character from those I have described. The continental rivers are large, and often occupy a wide-spreading bed, with numerous channels and islands, and in dealing with them, the main object of the engineer is to confine the whole of the water into one stream, in the certainty that, although the currents may be inconveniently *strong*, there will be no difficulty in securing ample *depth* for navigation.

The works lately undertaken on the Rhine, as described in the *Proceedings of the Institution of Civil Engineers*[1] by Mr. Jackson, are of great extent, and were attended with much difficulty on account of the sudden floods, which, he states, occasionally in the course of 24 hours rise to the extent of 38, and even 40 feet. They consisted of cuts for straightening the channel—dams or

Marginal notes: Improvement of upper portions of Continental rivers. Rhine.

[1] Vol. vii. p. 211.

weirs for checking the flow of the river—what are called diversion arms for separating the navigable channel, and spurs for directing its course.   It is worthy of notice that all these diversion arms and spurs are constructed of bundles of fascines, fixed by piles, and weighted with earth and stones—the distinctive feature of continental river-works.

Danube.   A short notice of the Upper Danube, as given by Mr. Shepherd,[1] will serve to illustrate the nature of the works on such rivers.   Mr. Shepherd says that the navigation was greatly impeded by the river shifting its course after almost every flood.   Its channel was divided into numerous branches, and the main object of the improvements was to shut off these lateral branches, and to cause the river to flow in one central channel.   The spurs formed for this purpose were projected from both banks of the river at such angles as were most suitable to the line of bank and the flow of the stream, which, being always in the same direction, and not reversed by a flowing tide, is more easily directed and controlled.

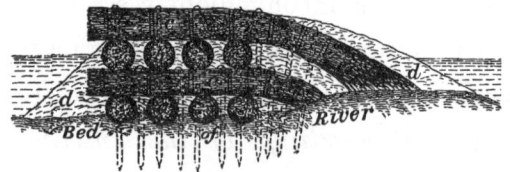

FIG. 25.

These spurs were constructed as shown in elevation fig. 25, and in plan fig. 26.   The series of fascines of brushwood, a and b, are bound or woven together so as

[1] *Civil Engineers' and Architects' Journal*, vol. xii. p. 321.

to form one continuous line throughout the entire length of the spur; each row is well secured to the ground by short piles or stakes, and the space between each row is filled in with earth.   The transverse fascines, *c c*, are laid

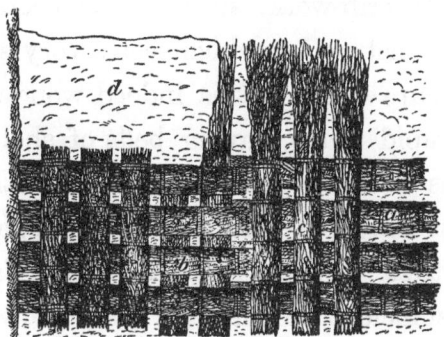

on the others, and also secured by piles—the ends towards the stream are left open, and their other ends are covered with earth, *d d*.   When the main spurs extending from the shore are completed, the arms of the river are dammed off in the same manner.

Mr. Shepherd says that the rapidity with which the spurs are made is truly surprising; they offer but little resistance to the water, and as soon as the sand-banks begin to move, the débris is deposited between and in the spurs, which renders them immoveable; the current at the same time is thrown into one channel, which has the effect of entirely scouring the river of the sand-banks previously deposited.

The brushwood is generally laid in such rivers in its green state, there to take root and grow again; consequently, after a few years, each of the spurs forms a thick massive hedge, which prevents the stream from making further ravages on its banks, and confines it to one central channel, scouring it out to a width and depth sufficient for all purposes of navigation.

The dams or weirs on the continental rivers are generally made of crib-work, such as has been described at page 146, a construction which seems very suitable where the currents are rapid. It is indeed wonderful what can be done in rapid rivers, having beds of shelving rock, by the judicious use of " crib-work " filled with stone. By cautiously sinking and loading these cribs, and gradually extending their dimensions, a structure is at last obtained, of sufficient weight and base to withstand the destructive action of the currents, even in the most rapid streams, the open spaces of the crib-work allowing the water to flow freely through until enough of stonework is deposited to secure stability. The boldest of such structures I have seen is the foot-bridge leading to Goat Island, across the rapids of the Niagara. The river at the spot is said to have a gradient of one in fifty-two, and the sight, as viewed from the bridge, of water tossed up by the rugged bottom into white-crested breakers, is, to an engineer, rather suggestive of a dangerous foundation. Niagara, with its falls and rapids, may truly be said to be unique, but as it is possible that the device adopted to form the piers of the bridge may offer hints available for some engineering purposes, I shall briefly describe the process as explained to me. The bridge was constructed by projecting trussed beams from the shore, balanced by weights at their inner ends. Upright supports were passed through holes in the outer ends of the beams, and driven down so as to rest on the bottom and serve as temporary supports. The beams thus temporarily placed, when planked, formed a gangway for the workmen, who commenced to sink at the end of the beams a small frame of open crib-work, which was filled with stones. This fabric, slender though it was, formed a more secure attachment for the ends of the trussed beams, so that heavy materials could be carried across them. Additional tiers of crib-work were successively built round the original core, until a mass was formed having sufficient base and weight to resist the force of the stream. Other piers were made in the same way, and thus the passage of the rapids was finally accomplished, the whole distance being about 400 feet. The only indication of the rude manner in which the bridge had been constructed appeared in the somewhat zigzag line of the roadway stretching from pier to pier, showing that the position of the piers had been fixed, not in accordance with any preconceived design, but to suit the rough surface of the river's bed, which, owing to the extreme rapidity of the current, could not previously be determined by soundings. Any attempt to sink a hollow box or cylinder, presenting a solid surface, in such a situation, would, I fear, be hopeless, but open crib-work, allowing the water to pass through while it was being gradually loaded with stone, solved the difficulty, even in the rapids of Niagara, and a similar mode of construction has been found very useful in erecting dams and similar works in many of the rapid continental rivers.

CHAPTER VII.

TIDAL PROPAGATION AND TIDAL CURRENTS OF RIVERS.

Importance of tidal flow—Its extent modified by circumstances—Tidal wave—Laws of its propagation—Tidal currents—Obstacles which operate in retarding tidal wave—Bore on the Dee—Bore on the Severn—Level of high water not raised by facilitating tidal propagation.

Importance of tidal flow.

TIDAL Navigation is a subject more intimately connected with the commercial interests of the British Isles, and occupies a more important position in the hydraulic engineering of this country, than those branches of river navigation which we have been considering in the preceding chapter. We have no great rivers here, like the Mississippi, on which to launch and navigate the largest class of shipping. Our fresh-water streams, even by all the aid of weirs and locks, can hardly be made deep enough for canal barges. The hills and valleys of our insular country have not sufficient area to form fresh-water streams available for navigation on a large scale.

The amount of water which our rivers discharge varies as the rain floods rise and fall; and, even at their best, our navigable rivers and estuaries may be regarded simply as creeks or inlets, formed and kept open, not by the fresh-water stream alone, but mainly by the action of the tide, and may be said to be navigable only when their channels are filled by the influx of water from the ocean. The

great agent in keeping open and deepening our navigations is to be found in the tidal flow, which not only
scours and maintains the sea-channels of our rivers, but
also does important service by increasing the scanty depth
of water which the river affords.   Nor is this all : another
most important advantage derived from the tides is that
upward current due to the tidal rise, which, at first
checking, and ultimately overpowering and reversing the
flow of the ebb-stream, carries vessels from the sea to
their inland ports without the aid of either steam or
wind.   Even the most superficial observer cannot fail
to recognise the value of this when he sees, on the Thames
or any other tidal river, a vast fleet of vessels of all sizes,
and from all countries, hurried on by the silent but powerful energy of the flowing tide.   How invaluable is such
an agent to the commercial interests of this country ! If,
indeed, the action of our river-tides were suspended, and
our coast begirt with a *constant* low water, with all its
attendant mud-banks and shoals, it might truly be said
of the steam-power employed in our factories and on our
railways, that its occupation would be gone.   I do not,
indeed, require to do more to enforce the wide-spread
interest of the subject than remind the reader that the
ports of London, Liverpool, and Glasgow, not to name
less important places, are entirely dependent on tidal
navigation for their existence.

From what has been said in Chapter III. as to the Its extent
physical boundaries of rivers, it will be apparent that the modified by circumstances.
extent to which this tidal influence is felt varies in different
situations.   Where the inclination of the river's bed is

gentle, and the channel is comparatively clear and unobstructed, it is felt far up the river, as in the case of the
Thames, where it reaches Teddington Weir, 65 miles from
the Nore ; and in the Tay, where it reaches its junction
with the Almond, 35 miles from the bar.   In other cases,
such as the Lune in Lancashire, or Dee in Cheshire,
the tidal flow is suddenly checked by artificial weirs
erected in the bed of the river for the use of mills.
In a third class of rivers the upward flow of the tide is
almost neutralized by the existence of natural obstructions, as in the case of the Erne at Ballyshannon, where
it flows only about three, and the Ness, where it flows
only about two miles up the river.

The tidal flow through estuaries and rivers gives rise
to two phenomena, the one called "tidal propagation,"
and the other "tidal current."   It is essential that the
difference between them should be clearly understood;
and it is further necessary that neither of them should
be confounded with that vertical rise and fall of the water
which is known as the *range* of the tide.

Tidal Propagation.

The tidal wave.   The tidal wave which enters an estuary is a branch of
the great tidal wave of the ocean.   Mr. Scott Russell was
the first experimental inquirer who conducted investigations on the tide wave of estuaries, and gave the laws of
its propagation, as deduced from experiments made on
the Dee in Cheshire, and the Clyde, which may be stated
as follows :—

Laws of its
propagation.   1. The great primary wave of translation differs from

every other species of wave, in its origin, its phenomena, and its laws.

2. The tide wave is *identical* with the great primary wave of translation.

3. In a rectangular channel, the velocity with which the tidal wave is propagated is equal to the velocity acquired by a heavy body falling freely by gravity through a height equal to half the depth of the fluid, reckoned from the top of the wave to the bottom of the channel. In a sloping or triangular channel the velocity is that of a gravitating body due to $\frac{1}{3}$d of the greatest depth. In a parabolic channel the velocity is that due to $\frac{3}{8}$ths or $\frac{3}{10}$ths of the greatest depth, according as the channel is convex or concave. And generally, *the velocity is that due to gravity, acting through a height equal to the depth of the centre of gravity of the transverse section of the channel below the surface of the fluid.*

4. The velocity in channels of uniform depth is independent of their breadth.

5. A tidal *bore* is formed when the water is so shallow that the first waves of flood move with a velocity so much less than that due to the succeeding parts of the tidal wave as to be overtaken by the subsequent parts, or whenever the tide rises so rapidly that the height of the first wave of the tide exceeds the depth of water at that place.

6. A wave of high water of spring-tides travels faster than a wave of high water of neap-tides.

7. In addition to these laws stated by Mr. Russell, I have found that the inclination of a river's bed, and of the

surface of the low-water stream, affect the rate of propagation independently of the form or depth of the channel,
and that, under certain conditions, a *decrease* of inclination is followed by an *acceleration*, and an *increase* of
inclination by a *retardation* of the rate of propagation.
But the results of my investigations will be best understood after I have described the works which afforded the
data on which they are founded, and I shall defer further
notice of them to a succeeding chapter.

These laws, stated by Mr. Russell, are supposed to
apply to the passage of the wave through channels having
a pretty uniform depth and form of cross section ; but the
very irregular outline of the beds of most of our tidal
channels renders it almost always difficult, and in many
cases impossible, to apply them rigidly to cases which
occur in actual practice.   I may, however, state generally,
in corroboration of the correctness of Mr. Russell's deductions, that after investigating the tidal phenomena of many
estuaries and rivers, I have found that in all cases the
quickest propagation of the tidal wave occurs at those
places where there is the greatest average depth; but
the varying outline of the cross section renders it almost
impossible in most cases to determine what is the *ruling
depth* for calculating the rates of propagation in any
particular section of the river.   In the Dornoch Firth, to
which I have already alluded, I found that the distance
of 11 miles between Portmahomac and Meikleferry is traversed by the tide wave in 30 minutes, being the interval
between the first appearance of the tide at the two stations,
giving a velocity of 22 miles per hour.   The depth of

water of that part of the firth varies from 9 to 50 feet. Between Meikleferry, again, and the Quarry, a distance of 8 miles, where the depth is much less, varying from 6 to 20 feet, the transit of the wave occupies 65 minutes, giving a speed of 6·4 miles per hour. Between the Quarry and Bonar Bridge, a distance of 1 mile, the water is comparatively shallow, varying from 1 to 3 feet, and the rise in the bed of the river is very rapid. In consequence of these obstructions the tide does not appear at Bonar Bridge for an hour and a half after it has appeared at the Quarry, giving a rate of propagation of only two-thirds of a mile per hour.

Tidal Currents.

But this passage of the tidal wave through an estuary or river must not be mistaken for the other phenomenon to which I have alluded, called the " tide current," which is totally distinct in its origin and character. The tidal wave which I have been describing as passing through the lower part of the Dornoch Firth, for example, at the rate of 22 miles per hour, is not that current due to the flowing tide by which vessels are carried across the bar, and borne onward to their destination. That current at the Dornoch Firth flows with a velocity which I never found to exceed 4 miles per hour. The laws of the propagation of the tidal waves, to which I first alluded, depend, as explained, on circumstances somewhat obscure ; but the velocity of the tide current, or that current which flows into our rivers, is due *entirely* to the *slope* or

fall on the surface of the water. The amount of this slope
depends on the rapidity with which the tide rises and the
degree of obstruction presented to its propagation up the
river. The more rapid the rise of tide and the greater
the obstruction to its flow the greater will be the difference of level in the tidal lines, as shown in the plates which
illustrate Chapter IV. A head of water is thus formed,
whose height is due to the rapidity of the rise of the tide
and the obstruction to its progress; and a flow of water
having a velocity due to that head is generated *up*, or
perhaps I should say *into*, the river or estuary, and this
flow of water is what I term the *flood-tide current*. A
similar slope or fall occurs on the surface of the ebbing
tide, due to the depression of the level of the sea at the
mouth of the river, which again causes an *ebb tidal
current*, flowing in the opposite direction.

Obstacles which operate in retarding tidal wave.

Now, the obstructions which are most frequently found
to retard tidal propagation, and to produce a heaping up
of the water and rapid tide currents, are the circuitous
routes of the channels of rivers, inequalities in their beds,
the projection of obstacles from their banks, and in certain
circumstances the slopes of their surfaces. The combined
effect of these obstructions is such as in all rivers to check
the propagation of the tide-wave, and, in situations where
there is a great and rapid rise of tide, to heap up the
water in the lower part of the river during flood, and so
to occasion what are termed " bores," and other apparent
anomalies. In the chapter on Tide Observations I
directed attention very fully to the existence of this heaping up of the tide during flood, but I did not then direct

attention to the cause and consequences, of which I have now to speak.   In the Dee, as has already been stated, there is at low water a fall of 11 feet from Chester to Flint, a distance of 12 miles; and on one occasion I found that after the tide had risen 18 feet 4 inches at Flint, it had not commenced to flow at Chester.   While, therefore, at low water there is a fall seawards of 11 feet from Chester to Flint, there was at the time alluded to a fall from the sea downwards, so to speak, of no less than 7 feet 4 inches from Flint to Chester.   Fig. 27 is a diagram

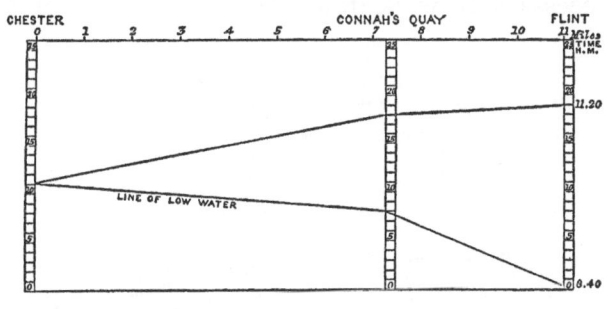

Fig. 27.

of these tide lines, which will illustrate more clearly the effect of this heaping up of water in the seaward part of the river.   The lower line represents the surface of low water, and the upper line shows the surface at the period of flood-tide to which I have alluded.   In this case the small depth of water, and tortuous and unequal channel, retarded the early waves of flood-tide so much, that they were overtaken by the succeeding waves; and, in accordance with Mr. Russell's theory, a *tidal bore* was the result, or, in other words, the water was heaped up so high, and the slope was consequently so great, as to cause the water to tumble over, and ascend the river in the form of a breaking wave.

Example of a
tidal bore on
the Dee.

The manner in which such tides flow up an estuary may probably be made more intelligible by a simple description of a flood-tide on the Dee than by diagrams of tidal lines. In fig. 28 the letters *a, b, c, d* represent a part of the low-water channel of the river Dee, at a place where the estuary is about 3 miles wide, and consists of extensive sand-banks. In examining minutely the windings of the stream in reference to certain investigations, it was necessary to walk down the right bank of the river

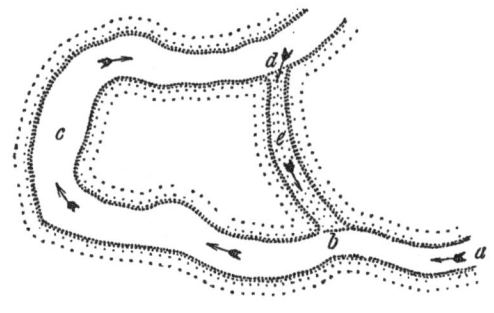

Fig. 28.

at low water, close to the edge of the channel. While so engaged, I crossed at the point *b*, a hollow in the sand-bank, which, though depressed below the general height of the surrounding surface, was nevertheless quite dry, the lowest part of the track being considerably above the level of the water of the river. Crossing this hollow, the noise of the approaching tide was heard; and expecting to meet the flood forcing its way up the river, I continued to walk on towards *c*; but seeing no appearance of its approach by the proper channel, and still hearing the noise gradually increasing, and apparently coming from behind, I turned round and perceived a rapid run of water flowing (in the direction shown by the arrow) through the hollow $d\,e\,b$, which had just been crossed, and emptying itself into the river at b. I immediately hastened back, and after having waded through the newly-formed stream at b, which had attained a depth of 6 or 8 inches, I remained on its upper side to see the result of this un-expected inroad. The water continued to rush through the hollow, rapidly gaining breadth and depth, and at last, after an interval of 2 or $2\frac{1}{2}$ minutes from the time at which the noise was first heard, the tide appeared forcing its way up the proper channel of the river, with a head or bore of 6 or 8 inches in height.   In this case it is clear, from what has been said as to the slope on the river from Flint to Chester during the early periods of tide, that the level of the water at d in the diagram would be above that at b.   The tide, on arriving at the point d, would be naturally divided into two branches or currents, one proceeding up the natural channel towards c, and the other flowing into the hollow in the sand-bank at d towards e; and as the level of the water at d rose, the stream which flowed into the hollow in the sand-bank would gradually rise higher until it sur-mounted the summit-level at e, after which it would rush from e to b without obstruction.   The other branch of the tide would in the meantime be forcing its way along the circuitous channel $d\,c\,b$, which was about a mile in length ; and before it reached b, the water at d had attained a much higher level than at b, and having surmounted the summit-level of the sand-bank at *e* continued to flow without obstruction into the channel of the river in the manner represented. From this I draw the general conclusion, *that in all places where the retarding influences which exist in the regular channel of a river exceed the obstructions in any back lake or swash-way, the tide will flow sooner through the latter than the former, and give rise to an apparent anomaly of two tides flowing in opposite directions, in the same river, till they are neutralized by coming in contact.*

Bore on the Severn.

The late Admiral Beechey, in his *Remarks on the Tidal Phenomena of the River Severn*, published in 1851, gives the following interesting account of the bore on that river:—"The bore," he says, "is not dangerous to boats if afloat in the middle of the river; and it is the common practice up the Severn to row the boats out to the centre of the stream on the approach of the bore, and put their head to the wave; but if this precaution be not taken, and the boats are allowed to remain at the edge of the shore, they are liable to be swamped or stove, as the wave breaks with great violence along the banks as it proceeds; but towards the centre of the river, if the water be not very shallow, the wave is smooth and unbroken. Before the arrival of the bore, the stream runs down the river, and the altitude of the water at a distance from the sea is quite stationary; but on the arrival of the bore the water instantly rises according to the height of the breast of the wave, and the stream turns and follows the wave up the river, although it had but a few minutes before been running down at a rapid rate; and this change of

stream is effected without any breaking wave. When
there is a heavy fresh down the river, and the stream is
running at the rate of four or more miles an hour, the
upward stream hangs for several minutes after the bore
has passed, not being able to overcome at the moment the
impetus of the ebbing water ; but when it has once
turned upwards, it attains its maximum speed in the first
half hour of the tide.   When the reaches of the river are
straight, the bore travels evenly up the river, but at the
turnings it is thrown off towards the further side, where
it rises higher than in the straight reaches ; thence it
recoils and impinges upon the opposite shore, and so, like
a disturbed pendulum, it oscillates from side to side, and
only regains its steady course when the reaches lengthen.
The highest tide of the year rolled up the Severn on the
1st of December.   There was about 2 feet of water above
the ordinary summer-level in the river, and the morning
was calm and favourable to the phenomenon.   The stream
at low water ran down at the rate of $2\frac{1}{2}$ miles (geographical)
per hour, until the time when the bore came rolling up
the river with a breast from 5 to 6 feet high at the sides,
and 3 feet 6 inches in the centre.   The wave was glassy
smooth ; and as it advanced towards a spectator stationed
at Stonebench, a singular effect was produced by the distorted surface of the wave reflecting the rising sun, and
brilliantly illuminating the stems and branches of the
wood skirting the river as the bore passed along—an
effect which greatly enhanced the interest of the phenomenon, which is at all times an object of curiosity.   The
stream turned up the instant after the bore passed, and

ran at the rate of 3¾ miles per hour, which was about half the average rate of the bore, the speed of which varied from 12 to 7 miles per hour, averaging 8 between Stonebench and Gloucester." Admiral Beechey further says, "that the effect of a fresh, or a certain depth of water in the river, upon the advance of the bore, is remarkable. At dry periods the great obstruction to the progress of the bore lies between Sharpness and Bollowpool, and at such times the many dry sand-banks prevent the bore attaining a rate greater than about 4 miles an hour; but when the river is under the influence of freshes, and the water raised, covering some of the banks, it appears to roll on at a rate of 10 miles an hour in opposition to the stream, which runs down at the rate of upwards of 4 miles an hour."

The phenomenon of the bore, called *mascaret* by the French, and in South America *pororoca*, has been reported by some travellers to assume dimensions hardly conceivable. Condamine, in describing that of the Amazon, says,[1] " At the distance of a league or two a frightful noise is distinguished, the herald of the pororoca, which is the name given by the Americans of the district to this tremendous *bore*. In proportion as it advances the noise increases, and shortly a promontory of water is seen, from 12 to 15 feet high, which is succeeded by a second, afterwards another, and sometimes again a fourth, rapidly impelled one after the other, and filling the whole breadth of the channel; this bore advances with prodigious rapidity, and carries away before it whatever opposes resistance."

[1] Pinkerton, *Travels*, vol. xiv. p. 252.

But to return to our own rivers: the object of the engineer in dealing with what I have termed the tidal compartment, is to facilitate the propagation of the tidal wave through the estuary or river, for which he has to design works, so as to *increase* the tidal influence, and also to *decrease* the tendency to the heaping up of water in the lower reaches of the river during flood-tide.   The heaping up of the tides, and the bore it occasions, will readily be admitted to be a great evil, and if my remarks as to our rivers being navigable only when they are supplemented by the presence of the tide be true, it will be no less obvious that all extensions of the period during which the tide is operative must be a great advantage; and this is what I mean by increasing the *duration* or *influence* of the tide.   It will be found, I think, in the examples I have hereafter to offer, that, with proper management, these desirable results may be surely accomplished, and the amount accurately determined.

This is probably the most convenient place to notice some facts of great importance in River Engineering, which I deduce from these considerations as to the nature of the tidal propagation and tide currents.   The obstructions to which I have alluded *retard* the rate of propagation, but they *increase* the velocity of the tide currents.   Now, as the aim, and, if successful, the effect of all engineering works, is to increase the rate of tidal propagation, no less certainly will they tend to lessen the heaping up of water in the lower reaches, and at the same time to *decrease the velocity of the tide currents.*   In cases where these currents are found to act prejudicially by producing a bore, or by bringing up sand from the lower parts of the estuary, or where they are inconveniently rapid for navigation, we are thus, while increasing the propagation of the tidal wave, enabled to check their energy, and thus to effect an important improvement.

The level of high water not raised by facilitating tidal propagation.

Another circumstance is worthy of notice. It is well known that the momentum of the column of water, flowing up the gradually contracting and rising channel of a river, causes the level of high water to stand higher than in the open ocean or in the lower reaches. This is accounted for, as already stated, on the principle of the conservation of forces. The height to which the water is thus raised depends on the quantity of water thrown in by the tide during a given time, the elevation being greatest at spring, and smallest at neap tides, as I have shown at page 83. At the Dee, for example, I found that the high water of spring-tides at Chester was 14 inches higher than that at Connah's Quay; while at neap-tides the difference of level was only 4 inches.[1] From observations given by Admiral Beechey it appears that the high water at Sharpness is sometimes 10 feet 8 inches higher than at Minehead, the distance between the places being 66 miles. But the rise of tide in the Bristol Channel is very great, and its tidal phenomena peculiar.

In considering the elevation of the level in the upper part of a river, as a mechanical question, Dr. Whewell says it may be accounted for by what is called " the

---

[1] Admiral Beechey found that at the Severn the low water of spring tides does not fall so low as that of neaps, which he attributes to the greater quantity of tidal water not having time to flow out. I am not aware of a similar observation having been made at any other place.

principle of the conservation of force.   When any quantity of matter is in motion, its motion is capable of carrying every particle of the mass to the height from which it must have fallen to acquire its velocity; but if the motion be employed in raising a smaller quantity of matter, it is capable of raising it to a height proportionally greater. In bays and channels, which narrow considerably, the quantity of water raised in the narrow part is less than in the wider, and thus the rise in such cases is greater."[1]

Now, as the effect of engineering works, as will be more fully detailed hereafter, is not only to produce a free propagation of the tide, but to admit a larger body of tidal water, it has been contended that such operations must necessarily cause the tide to rise higher, and it has been attempted to be shown in some cases in my own experience that they would necessarily occasion inconvenience, and even injury to property, by the improved river rushing up with violence and overflowing its banks. After the most careful observation, however, I have not been able to detect that such operations have, in any case, had the effect of notably raising the level of the high-water line.   The tide, in improved rivers, begins to flow earlier than before, and a larger body of water is carried up the navigable channel, where its effect is most useful, but the same works which *increased* the propagation have, by removing obstructions, *decreased* the heaping up of the tide, and, consequently, the velocity of the tide current; and by this fortunate compensative action, our rivers, though their beds are opened up and improved, do

----

[1] *Philosophical Transactions*, 1833, p. 204.

not inundate our towns, or even overflow our quays, but quietly keep within their original limits. That such improvements, by affording a more rapid discharge at ebb-tide and low water, have diminished the extent of land floods, cannot, I think, be doubted. At Glasgow, the floods do not now flood the Green and the low-lying streets in that locality as they used to do; and at the Tees, Mr. John Fowler, the engineer to the Navigation Trust, says that previous to the improvement of the river the lower parts of Stockton were frequently flooded, and in the High Street of Yarm, which is 8 miles above Stockton, the water often rose 5 or 6 feet, but since the execution of the works there has been no such flooding. But this land flooding, the reader will bear in mind, is in no way connected with the tidal phenomena which we have been considering. I have, however, noticed it, in passing, as an effect of river improvement.

CHAPTER VIII.

TIDAL COMPARTMENT—WORKS FOR ITS IMPROVEMENT.

Removal of obstructions to tidal flow—Weirs erected for public works—Works
for improvement of tidal compartment of rivers—1st, Removal of lateral obstructions; jetties objectionable; piers of bridges objectionable—2d, Training
walls; Ribble low-water training walls; comparative advantages of straight
and curved walls; form and construction of river walls—3d, Closing of subsidiary channels—4th, Substituting straight cuts for bends—5th, Dredging;
its introduction; bag and spoon dredge; bucket between two lighters; steam
dredges; hand dredges; dredging on the Clyde and Wear; improvements in
steam dredges; dredging on Amsterdam and Suez Canals; longitudinal and
cross dredging; blasting at Ballyshannon, at the Severn, and at St. Heliers,
Jersey; dredging in exposed situations—6th, Excavation; by diving-bell;
by floatation; by cofferdams—7th, Scouring—8th, Reducing the inclination
of the bed.

THE removal of all obstacles to the flow of the tide Removal of
is the object, as already stated, to which attention has obstructions
chiefly to be directed in designing improvements in the to tidal flow.
department of navigation we are now considering; and
in order to form a satisfactory opinion, it is necessary to
have an accurate survey, showing the depths of water and
the breadths of channel throughout the whole extent of
the river, as well as the amount of tidal range, the velocity
of the currents, the rise on the bed, and the nature of the
materials of which the bottom and banks are composed,
as explained in the chapter on Hydrometric Observations.
Possessed of this information, the engineer is in a position
to consider to what extent the bed of the river may, with
advantage, be deepened and widened, and the currents

directed by means of walls; also whether subsidiary
channels may, with safety, be shut up, or new cuts be
made for the passage of the river, or whether or not
irregularities in the width, which injuriously affect the
currents, may be corrected. The effect of such works
will be to cause the currents of flood and ebb tide to flow
always in one channel, and thus to exert their full and
combined power in keeping open one navigable track.

Weirs erected
for works.

Before, however, proceeding to describe these works,
it is perhaps proper to notice the artificial weirs that in
some tidal rivers have, in early times, been erected for
the purposes of manufacture. The removal of such erec-
tions, however prejudicial they may be to navigation, is
in many cases attended with difficulty, owing to the great
value of the interests involved, and the large compensation claimed by proprietors. Such weirs, for example,
are to be found on the Dee in Cheshire and the Lune in
Lancashire, and other rivers; and in order to show their
obstruction to the tidal flow, I have only to state the
facts as regards the weir on the Dee, which was erected
at an early date for supplying water-power for Chester
mills. I had occasion to examine it with reference to
the extensive flooding of the meadow-lands on the banks
of the river above the weir, and found that its crest was
11 feet 6 inches above the bed of the river immediately
below it, and that the level of the crest, if extended up
the river, does not strike the bed for a distance of 7
miles. It thus presents a perpendicular face to the flowing tide, which is completely checked until it rises so high
as to reach to the top of the weir, and this happens only at

high spring-tides. The weir, therefore, forms an artificial pond in the river's bed of 7 miles in length, which, were the obstacle removed, would be filled and emptied as the tide flowed and ebbed, and thus the scouring power of the river would be increased. But the vested interests of the mill-owners cannot be violated.

The removal of existing quays and other works of long standing, as in the case of the Tyne, the Wear, and other rivers, is also for the same reason difficult, and works must often be designed for such localities which shall not injuriously affect existing property, unless, indeed, as in the cases of the Thames between Westminster and London Bridge, and the Foyle at Londonderry, where the rights of all proprietors were purchased, and a line of quays formed to meet the public convenience irrespective of private interests.

These weirs and quays, however, present difficulties, which may be regarded as financial rather than engineering, but I have thought it right to notice them, in passing, and shall proceed to consider the works which will be found to be generally applicable to river improvements, under the following heads :—

1. Removal of lateral obstructions.
2. Training walls.
3. Closing of subsidiary channels.
4. Substituting straight cuts for bends.
5. Dredging.
6. Excavation.
7. Scouring.
8. Reducing the inclination of the bed.

1. *Removal of Lateral Obstructions.*

Under the "Removal of Lateral Obstructions" may
be classed all those works which have for their object the
formation of proper outlines for the banks or sides of the
Jetties
objectionable. river.   In the early history of river engineering it was
common to construct jetties or groins projecting from the
banks on either side, with the view of narrowing the
stream, and producing a greater scouring power to operate
on the bottom.   It is no doubt true that such projections
have the effect of producing a *local* acceleration of the
currents, and in soft bottoms a corresponding increase of
depth in their immediate vicinity.   But this increase of
velocity and depth being due entirely to the obstruction
and consequent raising of the level of the water caused
by the jetty, is strictly local.   Whenever the water passes
the end of the jetty, it expands into the greater width
of bed, the head is reduced, a stagnation or eddy takes
place, and a bank or shoal is formed—a result which
invariably follows the projection of any obstruction or
foreign body into a stream having a soft bottom.   I have

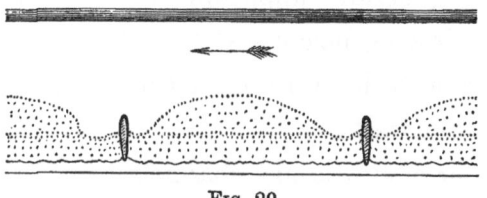

FIG. 29.

often observed this on the river Dee, in Cheshire, where
there is a long straight reach with jetties projecting from
one side, and a continuous embankment on the opposite
side of the channel, as represented by the dark lines in

fig. 29, which is a small portion of the Dee at low water, the direction of the current being shown by the arrow. From this cut it will be understood that at the ends of the two jetties there are holes 12 or 14 feet deep at low water, while a sand-bank or shallow, nearly dry at low water, extends into the river between the jetties. The manner in which the tidal current is distorted by the jetties is shown in fig. 30, which represents the same portion of the Dee during the flowing tide. The current, indicated

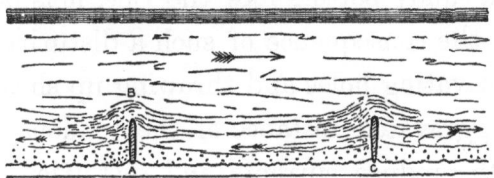

FIG. 30.

by the large arrow, on reaching the jetty A, is obstructed, and consequently the level of the surface is raised, so that the water on the seaward side of A is some inches higher than on its landward side. The surface of the river at B is also raised, and a strong current is generated past the end of the jetty, which curls round into the space between the jetties A and C. The main stream passing on, impinges against the jetty C, raising a head at its seaward side, and a corresponding current towards the shore C A, along which it flows, and toward the root of the jetty A, where the level of the water is also low. The whole of this complicated motion takes place in not very many seconds, and the result is a counter-current and eddy between the jetties ; and this action continues during the whole flow of the tide, and is more or less marked accord-

M

ing to the strength of the current; for it must be remembered that the depression on the upper sides of the jetties is not immediately filled up by the rush of water round the extremity, because the head which produces the velocity continues to increase with the rising tide, and the effect shown in fig. 30 will continue until slack tide, when the surface of the water above and below the jetties attains the same level. A similar action, the directions of the currents being reversed, takes place at ebb tide, and in a river with a sandy bottom like the Dee, it is not difficult to imagine the consequence of such a disturbance of currents in excavating holes and throwing up shoals.

As an aggravated instance of the tendency of all obstructions to produce currents and distortion of the bed of a river, I may refer to a vessel of about 170 tons, which, by the breaking of a tow-line, grounded in the

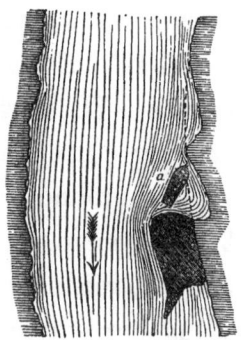

FIG. 31.

Tay when there was some flood in the river. The consequence of this mishap is shown in fig. 31, where the vessel is represented at a as lying in a pool which was scoured to the depth of 10 feet in the course of a few tides ; and the gravel thus excavated by the current, acting on the grounded vessel, and amounting to upwards of 1000 tons, was deposited in the form of a bank 5 feet above low water immediately below the pool, as shown in hatched lines. Thus, although the current in its natural state was not sufficiently strong to act on the bed of the river, the foreign body or obstruction caused by the grounded vessel raised a head of water which produced a current powerful enough to excavate hundreds of tons of gravel in a few hours. A similar effect, though varying in degree, occurs in all rivers confined by jetties, such as those on the Dee, to which I have referred. Rivers so treated present an alternation of shoals, nearly dry at low water, and deep pools instead of a regular bottom and a uniform depth of water available for the purposes of navigation ; and it is wonderful how long the system of "jettying," or, as it is termed in England, "cauling" a river continued to be advocated and followed by engineers. The Clyde, Tyne, Tees, and other rivers, suffered, if I may use the expression, from jetties, constructed at great expense, intended to confine and improve, but which rather tended to distort their channels. On the Clyde and some other rivers the ends of the jetties were latterly connected by longitudinal walls, which no doubt, to some extent, obviated the evils described as produced by the jetties on the Dee. But still it was long *generally* believed that jetties must, in the first instance, be constructed, even although their extremities might afterwards be connected with longitudinal walls if necessary. From the Clyde, the upper part of the Ribble, the

Tay, and the Tees, and other rivers they have been entirely removed.   I have universally found that, wherever jetties existed, their entire or partial removal formed one of the first steps towards an improvement of the navigation, being, in all cases which have come within my experience, followed by good results ; but I am not prepared to say that there may not be some special case in which a jetty may be advantageously employed in a navigable channel, indeed, I on one occasion recommended it, to reduce an undue width of channel, where circumstances connected with access to the land prevented the formation of a wall, and, of course, for other purposes, such as the protection of a river's banks without reference to navigation, they are often found to be very useful.

Piers of bridges objectionable.   What I have said as to the injurious effects of jetties applies with equal force to the piers of bridges erected across tidal rivers.

At the bridge over the Lary, near Plymouth, erected by Mr. J. M. Rendel, it was found that the scour was operating injuriously on the bottom, and he applied an artificial bed of clay from 18 inches to 2 feet thick, covered with stones of all sizes from 200 lbs. downwards, to protect the clay from the run of the water, the clay having a good effect in preventing the stones from being moved ; the combined thickness of the clay and stones was from 2 feet to 2 feet 6 inches, thus replacing the loss of the natural bed which had been scoured away. This artificial bed was found to be quite successful, and resisted a current of nearly 5 miles an hour.   Messrs. D.

and T. Stevenson suggested the same remedy for the railway bridge at Inverness, where scouring to the depth of 6 feet had occurred. In both cases, the original bed being restored, the velocity of the currents was increased by the reduction of water-way due to the piers of the bridge. But in navigable rivers it is not always desirable to increase the velocity of the current, but rather to found the piers so low as to place them beyond risk of injury from scouring. As an instance of this I may mention the railway bridge across the Foyle at Londonderry. The scouring which took place there operated chiefly on the eastern or concave side of the river, and the deepest cavities were in immediate proximity to the piers of the bridge,— results which, from what I have stated as to the Dee, would naturally happen. The greatest depth scoured from the bed of the river was about 4 feet between the piers, and increased at the piers themselves to about 8 feet. The transverse section of the river showed that the aggregate width of the piers of the bridge and side-walls was 150 feet, being about a sixth of the whole width of the river at the place in question. The sectional area of the river previous to the erection of the bridge was 18,079 square feet at half-tide, when the current is strongest, and the piers occupy a space of 2475 square feet, reducing the sectional area of the river by that amount. It is obvious that in all such cases, if an equal quantity of water is to flow into the upper part of the river in the same time as before the erection of a bridge, the velocity of the currents must necessarily be increased in proportion as the sectional area is diminished.

This increased current, acting on the soft bottom, must gradually gain in depth what has been taken from the water-way in width, and whenever the deepening is sufficient to compensate for the diminution of sectional area caused by the piers, the scouring action will cease. The sectional area occupied by the piers of the Foyle Bridge at half-tide was, as already stated, 2475 feet, and the sectional area scoured from the bed of the river was about 1868 square feet, and in reporting on the subject to the Harbour Commissioners it was stated that the scouring might be expected to continue until the increased sectional area should have so far diminished the velocity of the currents as to render them inoperative on the bed of the river—a result which has since been verified; and as the piers of the bridge had been carried by Mr. Hawkshaw, who was engineer to the railway company, upwards of 30 feet below the bed of the river, the stability of the structure was not affected, and the velocity of the tides through the opening draw of the bridge, made for the passage of ships, was not increased. A similar result happened at the railway bridge across the river Tay, where, from sections made in 1847 and 1849, before and after the bridge was built, it appeared that the bed of the river, which consisted of gravel, had been scoured to the extent of 2 or 3 feet, and that it ceased whenever the normal proportions between the quantity of water passed and the sectional area were restored. In a river like the Tay, where the bottom is gravel, it may fairly be assumed that when the scour of high floods has enlarged the water-way the bed of the river will continue unaltered, because the material that has been scoured out is so heavy that only the velocity of a heavy *land-flood* can move it, and, therefore, the mere flow of the tide cannot again disturb it. But in rivers having soft beds, composed of fine sand or silt, easily moved, the material scoured by land-floods is brought back by the flood, and thus it is impossible, when piers or other obstructions, such as jetties, are placed in rivers with soft bottoms, to preserve a uniform depth of water, as it changes from fortnight to fortnight with the constantly changing velocities of neap and spring tides.

The only other remark which I have to offer on this section of works is, that in some instances where the river is contracted by the projection of quays, or by the natural formation of the banks, it may be found advisable, where it can be done consistently with existing interests, to enlarge the cross-sectional area in order to reduce the velocity of the currents, and prevent disturbance of the tidal flow. Indeed, it may be laid down as a general principle in designing works for river improvements, *on the one hand, not to adopt a water-way so great as to reduce the scouring power and produce shoaling, nor, on the other hand, so small as to increase the current beyond what is convenient for the proper management of vessels.*

2. *Training Walls.*

In open estuaries filled with sand-banks, the courses of rivers are liable to constant alteration, due to every change in the tides or winds. The woodcut of the Lune, fig. 32, illustrates this remark. The several dotted lines represent the variations of the river during the period of

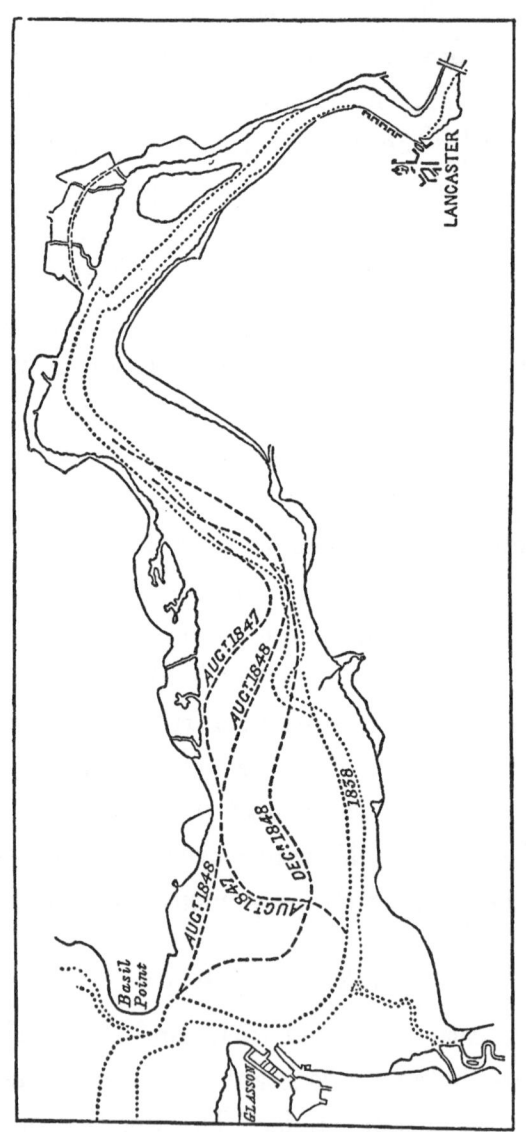

Fig. 32.

a few years. This tendency to deviate from channel to channel is common to all rivers that are left, undirected, to work their way through a tract of sand, and is utterly destructive to navigation. Continued during every flood and ebb tide it. sets loose a large amount of floating sand, which is daily drifted to and fro, and deposited in some new situation. A channel which is thus constantly shifting its course never remains sufficiently long in one position to form for itself a properly defined bed, but is in fact always in a transition state ; the sand which is worn from the *concave* side, where there is the greatest scour, being thrown to the *convex* side of the stream, while some portion of the floating materials, carried to and fro during this process of perpetual change, is often deposited, and forms shoals in the middle of the fairway. A river left in this state of nature cannot possibly attain the *maximum* depth due to the natural scour of the tidal currents, as their power is expended in abrading and removing the sand-banks through which the stream flows, and not, as it ought to be, in deepening and scouring its bed. In such cases what is wanted is to secure a permanent channel, by guiding the first of the flood and the last of the ebb tide by means of walls, so that the strength of the currents may constantly operate on the *same line of channel.* In this way it is obvious that not only will the advantage of a permanent navigable track be obtained, but the constant action of the currents of flood and ebb tide flowing in the same channel, will secure a much greater permanent depth than they could possibly do if permitted to wander at random through the estuary, sometimes operating in the same channel, and at other times directly opposed to each other, and these results can best be attained by training walls.

Ribble low-
water training
walls.
In 1836, Mr. James Walker was, I believe, the first to propose low parallel walls for the Clyde at Dumbarton; but I believe I am also correct in saying that the Ribble was the first river, passing through an estuary of sand-banks, which was improved by excavation and low training walls *alone;* and even long subsequent to the successful construction of training walls on the Ribble, jetties continued to be erected on the Tyne, the Tees, and other rivers. Even in 1850, twelve years after the Ribble works were commenced, I gave evidence in support of a Conservancy Bill for the river Tyne, which was intended to introduce a new system of treating the river; and in speaking of the condition of the Tyne at that time, I stated that " the works which have been executed to improve its condition, consisting of groynes or jetties raised above the level of high-water mark, and extending into the channel, are by no means judicious; that had such works been adopted as have of late years been introduced with unquestionable advantage in various rivers, the Tyne would, like them, have been in a condition very different from what it now presents; and moreover, by the present system of working, I believe that the Tyne, viewed as a whole, *never can be improved to any great extent,* if at all." I need not add that these remarks in no respect apply to the present state of the Tyne navigation.

The proposal to improve the river Ribble by boldly projecting rubble walls through its sandy estuary, without any cross jetties to check the tide and prevent its flowing up on either side behind the walls, was considered as hopeless, and it was with no little difficulty the directors of the Company, amid much contending local advice, could be persuaded to try the experiment; and it was not until after several interviews that the Admiralty, represented by Admiral Sir F. Beaufort, as hydrographer, and Captain Washington, as chief officer in the hydrographic department, gave the official assent to works which have since, in many instances, proved to be the proper treatment for such a river.

Questions have been raised as to the comparative advantages of straight and curved walls for directing a channel. I believe the engineer will find that, in most cases, the direction of such walls must be determined, not by any abstract consideration as to the superiority of straight or curved walls, but chiefly by the relative positions of the points between which the stream is to be conducted, and the outline and geological formation of the shores and banks of the estuary that intervene between those points. The consideration of such matters may render it expedient, according to the special circumstances of the locality, to adopt walls having *concave, straight,* or *convex* outlines, as shown in figs. 33, 34, and 35.

Viewed as a purely abstract question, it may, I think, be safely affirmed that a stream is most likely to follow a permanent course when directed by a concave wall, as shown in fig. 33, in which the axis of the stream is represented by the dotted line. Dr. Young observes that the

Comparative advantages of straight and curved walls..

centrifugal force in curved channels has a tendency to draw the greater portion of the water to the concave side, and thus the greatest scouring power, and consequently the greatest depth of the stream, will be found upon that side, as must have been observed by all who have had occasion to study the subject. In a channel directed by straight walls (fig. 34), the current has no such decided bias for either wall, and is consequently more easily thrown across from side to side. A wall, on the other hand, having a convex outline, as shown in fig. 35, is

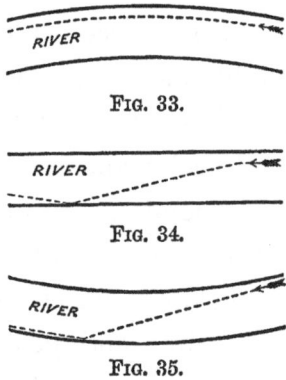

FIG. 33.

FIG. 34.

FIG. 35.

(especially if the radius of curvature be small) still less suitable as a guide, as the line of wall diverges from the direction of the axis of the current. These remarks are not hypothetical, as I have found that their correctness has been verified by cases in actual practice. There is doubtless some disadvantage in the deep water being on one side of the channel, as shown in the cross section, fig. 36. It would be more convenient for navigation were the deep water in the centre; but it is found that the current has a tendency to adhere to one or other of the walls, and it is better that the channel should keep constantly to one wall, than that it should alternate from side to side, as is more apt to be the case in absolutely straight channels. It is, however, proper to state that Mr. Fowler, the engineer of the Tees Navigation, has found no practical difficulty in maintaining very constantly

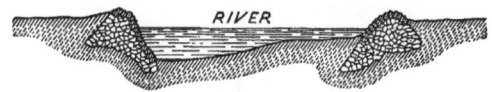

Fig. 36.

a fair navigable channel in the long straight reach on the Tees, upwards of a mile in length, where the river is trained by two parallel walls.

The direction of river walls must, however, be carefully considered by the engineer with reference to existing circumstances, as it is a point which clearly requires that every case be judged on its own merits. But I think it will be found safe in executing such works to adhere as closely as possible to the following general rules, which are the result of experience :—

First, The channel through open estuaries should, in all cases where funds will admit of it, be guided by double walls. In cases, however, where the estuary is bounded by a hard beach, presenting a favourable line of direction, a single wall may occasionally be found sufficient. All curves which it may be necessary to introduce should be of as large a radius as possible, and should, if practicable, be tangential to each other, or to the straight parts of the line with which they are connected.

Second, The walls should not be *raised* to a higher level above the low-water line than is absolutely neces-

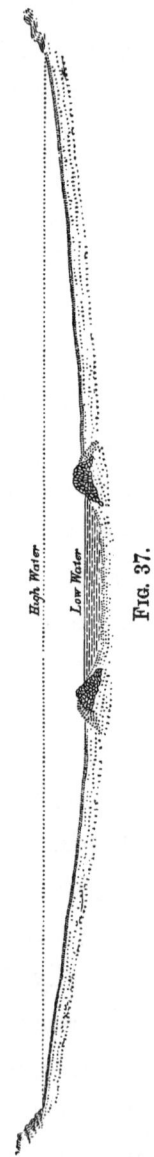

High Water

Low Water

FIG. 37.

sary for the purpose of conducting the early and late currents of the tide; and their direction should be marked by occasional perches.

Fig. 37 represents the disposition of such walls in estuaries, as executed under the direction of Messrs. Stevenson. They are raised from 3 to 5 feet above the low-water line, so that, while they guide the low-water channel, they do not prevent the tide at high water from flowing on either side of them and filling the estuary.

Third, River walls should, during their erection, be pushed forward with vigour, and not in a desultory, timid manner; the effect of such a course being to increase the depth of water in which the wall has to be made, and the amount of stone required for its construction.

Fourth, It will be found that such walls as I have been describing will be most advantageously formed of rough rubble stones, backed with clay and gravel, in the manner shown in fig. 38.

Fifth, In determining the proper width of channel to be formed by training walls, the engineer must be guided by a careful consideration of the fresh-water and tidal discharge, and the trade to be provided for. The walls on the lower part of the Ribble and the Tees are from 400 to 500 feet apart.

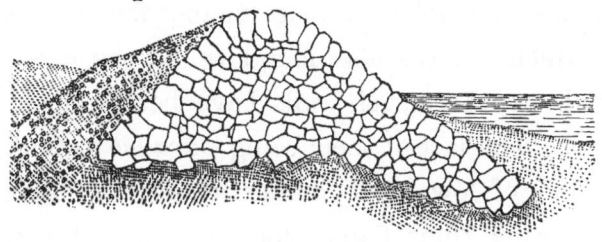

FIG. 38.

Care must be taken not to adopt a channel of too great width, so as to decrease too much the scouring power of the currents. It was found on the Dee, from a series of observations kept for eight years, that the two places, at which the depth was most frequently below the standard of 15 feet at high water, were Saltney, where the channel had been increased in width to accommodate shipping, and Upper Ferry, where it had been enlarged to suit the ferry traffic. Out of 139 instances of deficient depth in the river, during the eight years, 65 had occurred at Upper Ferry and 38 at Saltney. It was found at Saltney that the vessels moored at Saltney wharves had the effect of increasing the scour on the bottom, and keeping the deep water close to the line of quays. An interesting result of the eight years' observations was, that the navigation between Chester and Connah's Quay, on *the whole*, was deepest in February after the winter floods, and shallowest in September and October, which accords pretty much with my experience of similar rivers, and will, I believe, be found to be of *general application* to all British rivers.

It was found by Mr. Park, under whose immediate directions, as local engineer, the walls on the river Ribble were constructed, that their foundations, with few exceptions, did not sink more than a few feet below the sand, which I have also found to be the case in many other places where such walls have been formed in sandy bottoms. The fact of their being *parallel* to the tidal currents, and of low level, prevents any very severe scour from affecting their foundations, as the water, so soon as it overtops the wall, gets relief by flowing over their banks on either side of the channel; but notwithstanding this flow of the current over the walls, the track between them, being the deepest part of the river, will be found always to afford the best navigable current, so that in navigating the channel there is no tendency for vessels to be sent across the walls; indeed, I am not aware of a single instance of ships fouling low training walls. In designing walls for the lower part of the Dee, at Connah's Quay, Telford, with a view, no doubt, chiefly to making land, advised the River Dee Land Company to make their bank much too high, and combined with it the old transverse jetties. The wall at Connah's Quay is carried up nearly to high-water level, and the scour was so great that an unnecessarily large amount of stones was expended in filling up the deep pools scoured below the level of the sand. In designing improvements for the Dee Navigation, in 1840, I recommended that the high-level walls should be discontinued, and that a wall on a low level should be adopted, similar to the works executed on the Ribble, where considerable difficulty was encountered in introducing the low parallel walls, as explained at page 186. In forming these walls it will be found necessary from time to time to add additional stones to make up slips, before attempting to pitch the top or face of the wall.

3. *Closing Subsidiary Channels.*

The next work to be noticed is the closing of what I term subsidiary channels, which are sometimes called

N

back lakes or *blind channels*, and are caused by islands in the river, so that instead of flowing in one broad, deep, and navigable bed, kept open by the whole available scouring power, the river is divided into two shallow channels, neither of them affording a good navigation, while frequently a ford or shallow is deposited both above and below the island caused by the disturbance which occurs at the junction of the divided currents. The object of these operations will be understood from fig. 39, which shows a portion of the Tay. On the Tay

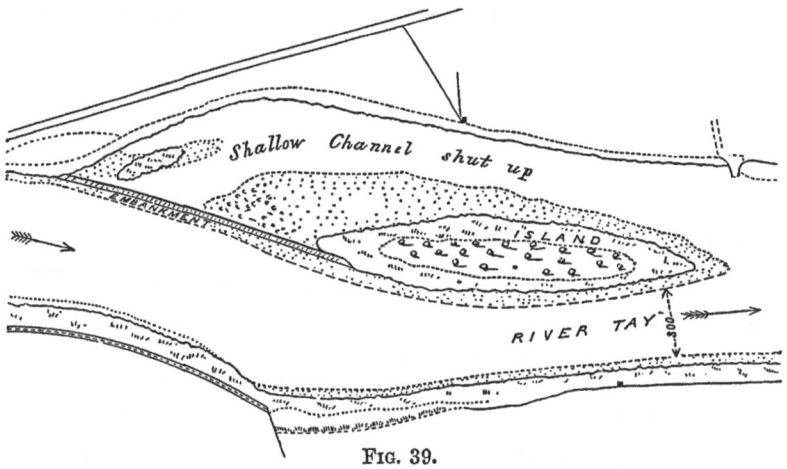

FIG. 39.

and the Lune several such secondary branches were, with much advantage to the navigation, closed up by means of embankments, formed of gravel dredged from the river, while the other or principal branch was enlarged and deepened, so as fully to compensate for the closing of the smaller channel, and assimilate its cross-sectional area to the rest of the navigable track.

The closing of the streams behind the islands on the Tay, and turning the whole flow into one channel, was not followed by a *sensible* rise in the surface or *flooding of the shores* in the augmented state of the river. But there could be no doubt as to the *increase* of its velocity, which was very marked. For some time, indeed, it was found that every flood wasted and cut the banks to such an extent as rendered it impossible to draw the fishing-nets until the sectional area had been enlarged, and the current so reduced as to admit of the banks being properly dressed and laid with broken stone, showing that there may not be any sensible increase of *height* in the surface, even when there is a considerable aug-mentation in the quantity of water discharged; but see-ing that the increased velocity is due to increased slope, there must have been some elevation of the surface opposite to the closed entrance, so that it is not possible, as averred by certain philosophers of the Italian school, that a small river may enter a large one without increas-ing its sectional area.[1] The shutting up of all such lateral branches should invariably be preceded by a careful com-parison of accurate sections of the two channels and the velocities of the currents running in them.

The embankments as shown in fig. 39 should be made at the upper end of the channel to be closed. They should be raised gradually across its whole width. They may be made of fascines or pile-work, when the bed is soft, and in cases where the dredgings from the river consist of heavy gravel, I have made banks by simply depositing the dredged materials across the mouth of the channel, allowing the currents to scatter them to the

[1] See Frisi's remarks on this controversy, p. 61.

proper slope. Dredgings so deposited from day to day will ultimately bring the bank to its proper height, when it can be further strengthened from behind, by floating the punts at high water up the old channel, which may be left open at the lower end so as to allow it to silt up.

4. *Substituting Straight Cuts for Bends.*

In rivers which follow a tortuous course navigation is sometimes greatly impeded by the abruptness of the bends, and the difficulty of navigating a ship through them, and where practicable it is sometimes desirable to obviate this by forming straight cuts. This is an operation, however, that must not be entered on without careful study of the tides and due consideration of the effect of the altered slope and currents on the river's bottom above and below the site of the cut. The levels should also be accurately determined and considered, as it may happen that the substitution of a straight cut for a long detour may involve a rate of inclination so steep as to induce currents injurious to the bed and banks of the river, and inconveniently rapid for navigation. Cuts have been formed with great advantage on some rivers, and I may particularly mention the Tees as a case where they have been successfully adopted. The Maudale Cut near Stockton was made, in 1810, and by a short course of 220 yards it cuts off a detour of nearly $2\frac{1}{2}$ miles, the navigation of which was exceedingly intricate. A second cut of 1100 yards in length was made on the Tees, in 1830, to cut off the Portrack bend, another inconvenient detour in the river's course.

Probably the largest artificial water-channel that has been executed, is that for the diversion of the river Dee at Aberdeen, for the particulars of which I am indebted to Mr. W. D. Cay, engineer to the Aberdeen Harbour Trustees. The Harbour Trustees, in the year 1868, obtained Parliamentary powers to divert the river, and to reclaim an area of land amounting to about 120 acres. Of this area it is intended to reserve a portion as water space to serve in the meantime as a tidal basin, and subsequently, when required, to be formed into a dock. The remainder will be embanked and formed into quays and streets, and partly feued for public works. The length of the new channel is 2000 yards, that of the old course being 2500 yards. The bottom is excavated to a uniform slope of

FIG. 40.

1 in 1300 downwards towards the sea, being parallel to the expected highest flood-line. The width of the channel at the bottom, as shown in fig. 40, is 170 feet; the bank on the north side has a slope of 3 to 1, and is protected with piles, clay, and stonework; on the south side the slope has an inclination of 10 to 1, this flat slope being arranged for the convenience of the salmon-fishings. The tops of the banks are 25 feet above the bottom of the channel, and the width at the level of the top of the banks is 495 feet. The bottom of the channel at the upper end is 2 feet, and at the lower end 6 feet below low water of ordinary spring-tides; high water of the same tide rises 12 feet 9 inches above the low-water level.

The amount of excavation of the new channel was about 1,036,000 cubic yards; about 916,000 cubic yards were taken out dry by manual labour, and drawn out in waggons, the remainder,—120,000 cubic yards,—being dredged.

5. *Dredging.*

The introduction of even the simplest of mechanical appliances for excavating materials under water, raising them to the surface, and depositing them in barges, was an important era in canal and river engineering. The first employment of machinery to effect this object is, curiously enough, like the discovery of the canal lock, claimed alike for Holland and Italy, in both of which countries dredging is believed to have been practised before it was introduced into Britain, and the moving power at first employed was, it need hardly be said, manual labour.

Bag and spoon dredge.

The Dutch, at a very early period, employed what is termed the "bag and spoon" dredge for cleaning their canals. It was simply a ring of iron, about 2 feet in diameter, flattened and steeled for about one-third of its circumference, having a bag of strong leather attached to it by leathern thongs. The ring and bag were fixed to a long pole, which, on being used, was lowered to the bottom from the end of a barge moored in the canal or river. A rope made fast to the iron ring was then wound up by a windlass placed at the other end of the barge, and the spoon was thus dragged along the bottom, and was guided in its progress by a man who held the pole. When the spoon reached the end of the barge where the windlass was placed, the winding was still continued, and the suspending rope being nearly perpendicular, the bag was raised to the surface, bringing with it the stuff excavated while it was being drawn along the bottom. The windlass being still wrought, the whole was raised to the gunwale of the barge, and the bag being emptied, was again hauled back to the opposite end of the barge, and lowered for another supply. This system is slow, and only adapted to a limited depth of water and a soft bottom. It has, however, been generally employed in canals, and much used in the Thames and at the Foss Dyke,

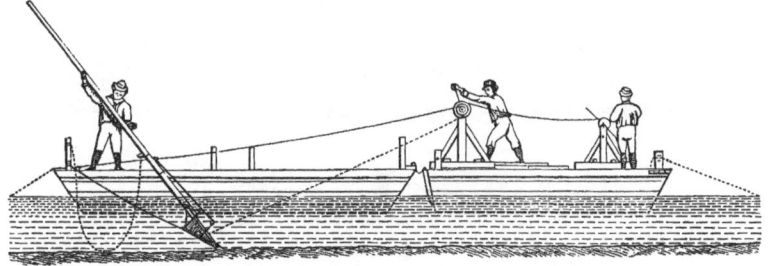

FIG. 41.

owing to want of space and other peculiarities. I found it could be usefully employed when the steam-dredge, though built expressly to suit the contracted limits within which it had to work, could not be used. The quantity raised at the Foss Dyke, by manual labour, in this way, was about 135,000 tons, and the cost did not exceed 7½d. per ton. Fig. 41 shows the manner in which the bag and spoon were employed.

Another plan, practised at an early period in rivers of considerable breadth, was to moor two large barges, one on either side; between them was slung an iron dredging bucket, which was attached to both barges by chains

Dredging by bucket between two lighters.

wound round the barrels of crab-winches worked by six men in the one barge, and a simple windlass, worked by two men, in the other. The bucket being lowered at the side of the barge carrying the windlass, was drawn by the crab-winch on the other barge across the bottom and up a sloping platform, which was lowered to the bed of the river, and was ultimately emptied in the barge. It was again lowered, and hauled across by the opposite windlass for a repetition of the process. This plan of dredging was adopted in the Tay till 1833, and fig. 42 will

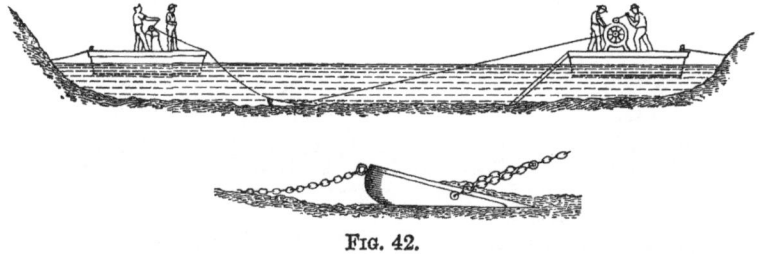

FIG. 42.

give a pretty good idea of the manner in which the apparatus was worked. The dredging-spoon is shown on a large scale at the foot of the cut.

These early *efforts*, as perhaps they may be called, at dredging, are, I think, worthy of being recorded, and will be interesting as compared with the more perfect machinery used in the present day, some of which I shall have to notice.

team dredges.   In all large operations, these and other primitive appliances have now, as is well known, been almost superseded by the steam dredge, which was first employed, it is believed, in deepening the Wear, at Sunderland, about the year 1796. The Sunderland machine was made for Mr.

Grimshaw by Boulton and Watt.[1] Receiving improvements from Mr. Hughes, Mr. Rennie, Mr. Jessop, and others, the steam dredge, as now generally constructed, is a most powerful machine in skilful hands, excavating and raising materials from the depths of 15 to 30 feet of water, according to the size of the machinery, at a cost not very different from, and in some cases even less than, that at which the same work could be performed on dry land.

As to the nature and extent of work that may be accomplished by dredging, I may state, generally, that almost all materials, excepting solid rock or very large boulders, may be dredged with ease. Loose gravel is probably the most favourable material to work in; but a powerful dredge will readily break up and raise indurated beds of gravel, clay, and boulders, and even find its way through the surface of soft rock, though it will not penetrate very far into it. In such cases it is usual to alternate on the " bucket-frame " a bucket of sheet-iron, for raising the stuff, with a rake or pronged instrument for disturbing the bottom.

Hand dredges have been used by Messrs. D. and T. Hand dredges. Stevenson at several places, by means of which even disintegrated or rotten rock has, at least to a limited depth, been raised; and I believe that in very many cases the surfaces of submerged rocks may, by means of such machines, be to a small extent broken up and removed, so as to obtain in certain situations a considerable increase of depth, without recourse to cofferdams, which involve

[1] *Encyclopædia of Civil Engineering*, by Edward Cressy, London, 1847; "The Dredging Machine," Weale's *Quarterly Papers*, Part i., London, 1843; *The Improvement of the Port of London*, by R. Dodd, Engineer, 1798.

great expense, as well as interruption to the traffic. These small dredges are worked by seven or twelve men, and cost about £230 to £390. They can work in a depth of about 16 feet, and can raise ordinary deposit at a cost of 1s. 6d. to 2s. per ton.

The construction of large river steam-dredges is now carried on by many engineering firms, each one naturally advocating his own arrangement of parts, and consequent superiority of performance.

Dredging on the Clyde.

For details as to the amount and cost of work done on a river where much dredging is annually performed, I perhaps cannot do better than refer to the Clyde, for in no river has dredging been more extensively or success-fully employed. I am indebted to the kindness of Mr. James Deas, the engineer to the Trustees of the Clyde Navigation, for the following information regarding the apparatus employed, and the extent and cost of the work done, which will be found both interesting and valuable.

Mr. Deas says truly that the Clyde Trustees employ probably the largest dredging fleet of any trust in the king-dom, in maintaining and still deepening and widening the river, to meet the ever increasing demands of the shipping trade.

Last year 904,104 cubic yards, or about 1,130,000 tons, were dredged from the river, of which 689,560 cubic yards were carried to sea by steam hopper barges, and 214,544 cubic yards deposited on land by means of punts. Of this 904,104 cubic yards, 345,209 cubic yards were de-posit from the higher reaches of the river and its tribu-taries, and from the city sewers, and 558,895 cubic yards new material. The total cost for dredging and depositing was £35,448, or about 9·41 pence per cubic yard.

Owing to the difference in power of the dredging machines employed, and the character of the material lifted, the cost of dredging varies much. Last year the most powerful machine, working 2420 hours, lifted 430,240 cubic yards of silt and sand, at a cost of 2·60 pence per yard, and this was deposited in Loch Long, 27 miles from Glasgow, by steam hopper barges, at 5·46 pence per yard. On the other hand, another dredger working 2605 hours, lifted only 26,720 cubic yards of hard gravel and boulder clay, at the cost of 20·8 pence per cubic yard, which was deposited on the alveus of the river, at the cost of 17·46 pence per cubic yard; another, working 1831¾ hours, lifted 122,664 cubic yards of silt, sand, and sewage deposit, at the cost of 5·67 pence per cubic yard, which was deposited on land at the cost of 16·40 pence per cubic yard; and another, working 2233 hours, lifted 65,160 cubic yards of till, gravel, and sand, at the cost of 5·89 pence per cubic yard, which was deposited on the alveus of the river at the cost of 9·83 pence per cubic yard.

The total quantity dredged from the river during the last twenty-seven years amounts to 13,617,000 cubic yards, or upwards of 17,000,000 tons.

The dredging plant of the Trustees comprises 6 steam dredgers, 14 steam hopper barges, 1 tug steamer, 3 diving-bells, 270 punts, and numerous row boats. The expenditure for wages of crews, coal, and stores, amounted last year to fully £14,000, and for repairs £10,775.

The value of the dredging plant employed is about £140,000.

Mr. Deas has also kindly furnished the following tables, from which the reader will see the gradual increase that has been made on the size of the dredging machines to meet the increased depth of water and growing necessity of increased accommodation for the larger class of vessels which now frequent the river :—

GENERAL DIMENSIONS OF DREDGERS EMPLOYED ON THE CLYDE IN 1872.

No.	Year built.	Length.	Breadth.	Depth.	H.P.	Greatest depth can dredge in.	Single or double bucket ladder.	Remarks.
		Ft. in.	Ft. in.	Ft. in.		Feet.		
1	1851	99 9	32 4	10 0	40	22½	Double	Punt Loading Machine.
5	1841	95 6	22 6	10 4	24	18	Single	Do.　　　Do.
6	1855	121 0	33 6	10 0	40	25	Double	Hopper Barge Do.
7	1860	108 6	23 6	9 0	25	25	Single	Punt　　　Do.
8	1865	161 0	29 0	10 0	75	28	Do.	Hopper Barge Do.
9	1871	161 0	29 0	10 0	75	30	Do.	Do.　　　Do.

No. 8 Dredger—
 Length, 161 feet.
 Breadth, moulded, 29 feet.
 Depth, 10 feet.
 Engine, 75 horse power.
  Cylinder, 48 inches diameter.
  Stroke, 3 feet.
 One bucket ladder, 90 feet 9 inches between centres.
 Size of buckets, 3 feet 3 inches × 2 feet 5 inches × 1 foot 11 inches.
 When working in sand, can lift 190 cubic yards per hour.
 Greatest depth can dredge in, 28 feet.
 Working draught, 6 to 7 feet.
 Wages, per day of 10 hours—Master, 7s. ; mate, 3s. 9d. ; engineer, 6s. 8d. ; firemen, 3s. 8d. ; assistant do. and cook, 3s. 4d. ; bow crabman, 3s. 4d. ; stern do., 3s. 4d. ; deck hands, three at 3s. 2d., one at 3s. ; watchman, 3s.

Coals, per day of 10 hours, 65 cwts.
Tallow,   „   „   2 lbs.
Oil (Lard), „   „   16 gills.
Waste,   „   „   1½ lbs.

Hopper Barge—

Length, 145 feet.
Breadth, moulded, 25 feet.
Depth, 11 feet 9 inches.
Engines, 40 horse power.
Draught, light (average), 5 feet 6 inches.
    „   loaded, 11 feet.
Speed, 8 to 9 miles per hour.
Cargo, 320 cubic yards, or say 400 tons.
Average distance run, loaded, 20 miles.
Wages, per day—Master, 7s.; mate, 4s. 6d.; engineer, 5s. 10d.;
    fireman, 3s. 6d.; deck hands, three at 3s. 4d.
Coals, per day of 10 hours, 70 cwts.
Tallow,   „   „   5 lbs.
Oil,   „   „   20 gills.
Waste,   „   „   2 lbs.

Quantity and cost of dredging done by No. 8 Dredger during year ending 30th June 1871 :—

							£	s	d
Wages,	.	.	.	.	.	.	678	0	0
Coals,	.	.	.	.	.	.	371	18	3
Stores,	.	.	.	.	.	.	182	7	1
							£1232	5	4
Repairs,	.	.	.	.	.		1669	6	11
							£2901	12	3

Interest and depreciation—
    Cost of dredger, £17,653, at 10 per cent.,   1765 6 0

                                £4666 18 3

Time worked during year, 2419¾ engine-hours.
Sand, silt, till, and gravel lifted, 430,240 cubic yards.

$$\frac{430,240}{24,193\frac{3}{4} \text{ hours,}} = 177 \cdot 80 \text{ cubic yards lifted per hour.}$$

$$\frac{£4666 \quad 18 \quad 3}{430,240 \text{ cubic yards,}} \quad 2 \cdot 60 \text{ pence cost per cubic yard lifted.}$$

Quantity and cost of conveying and discharging the total dredgings lifted by Nos. 6 and 8 Dredgers during the year ending 30th June 1871—

Wages, coals, and stores,	£6917	0	5	
Repairs, . . .	3255	7	9	
			£10,172	8 2

Interest and depreciation—
Cost of 10 Hopper Barges, £51,510, at
10 per cent., ... 5151 0 0

£15,323 8 2

$$\frac{£15,323 \quad 8 \quad 2}{673,240 \text{ cubic yards,}} = \frac{5\cdot46 \text{ pence cost per cubic yard.}}{\text{total dredgings conveyed.}}$$

Note.—Four Hopper barges are required to keep one dredger in constant work.

ABSTRACT of the Quantity and Cost per cubic yard of Dredging and Depositing during the year ending 30th June 1871.

Dredger, etc.	Nature of stuff, and where dredged generally.	Total cubic yards lifted.	Cubic yards lifted per engine-hour.	Pence per cubic yard.					
				Dredging.	Conveying and depositing by barges.	Towing punts to discharging ground.	Discharging dredgings from punts.	Punts, boats, etc., used.	Total cost.
No. 1 Dredger,	Sand, silt, and sewage from Glasgow harbour.	122,664	66·96	5·67	...	2·88	10·00*	3·52	22·07
No. 5 Do.	Hard till, gravel, and sand, from Erskine Ferry, etc.	65,160	29·18	5·89	...	1·76	5·42†	2·65	15·72
No. 6 Do.	Sand, clay, and mud, from Pt. Glasgow, etc.	243,000	83·19	3·36	5·46	...	...	...	8·82
No. 7 Do.	Hard till and clay from Erskine Ferry, Elderslie, etc.	26,720	10·26	20·81	...	2·34	5·42†	9·70	38·27
No. 8 Do.	Sand, silt, till, and gravel, from Glasgow and Bowling harbours, etc.	430,240	177·80	2·60	5·46	...	...	...	8·06
10 Hopper barges.	...	...	...	...	average 5·46				
Tug steamer,	...	...	...	...	...		average 2·83		

Nos. 1, 5, and 7 are punt-loading machines. Nos. 6 and 8 are hopper barge machines.

* Contractor's price for discharging at Blytheswood Park, *including slip docks, and waggoning a distance of about ¾ mile.*

† Discharging by Trustees' men on river banks near Erskine Ferry, *by beaching punts and wheeling.*

Mr. Murray has given me the following tabular view Dredging on the Wear. of the dredging of the Wear at Sunderland, which is also an interesting record of the quantity and cost of material raised by a dredging machine ; but this view is not given by way of comparison with the preceding, as there is no analogy between the cases. The contracted state of the Clyde, the frequent interruptions to which the work was subject by the constant passage of vessels, the expense of removing and depositing the stuff, and the higher rate of wages, must necessarily have increased the cost of executing the work in that situation.

TABULAR VIEW OF THE DREDGING OF THE WEAR AT SUNDERLAND IN 1842-46.

Date.	Total quantity raised per annum.	Expenditure in labour for raising and depositing per annum.	Expense of Fuel per annum.	Expenditure in Labour for Repairs per annum.	Expenditure in Materials for Repairs per annum.	Total Expenditure per annum.	Average cost per ton on the year's expenditure.
	Tons.	£ s. d.	£ s. d.	£ s. d.	£ s. d.	£ s. d.	Pence.
1842	128,245	922 1 2	111 0 0	754 16 0	704 17 11	2492 15 1	4·665
1843	141,325	879 16 0	70 0 0	503 13 4	786 13 11	2240 3 3	3·804
1844	90,980	567 13 4	66 5 9	259 2 1	563 9 10	1456 11 0	3·842
1845	101,075	721 9 0	66 7 6	336 8 0	527 7 10	1651 12 4	3·921
1846	140,350	724 5 4	58 2 5	500 17 2	520 3 2	1803 8 1	3·083

1842 to 1846	Hence the average cost per ton on five years' work—	
	For raising and depositing at sea, . . .	= 1·528
	For fuel, ...	= 0·149
	For labour in repairs, ...	= 0·943
	For materials to ditto, ...	= 1·243
	Average total Expenditure, . .	= 3·863

This is perhaps the best place to mention some im- Improvements in steam dredges. provements that have been suggested on the present, or I may, I suppose, term it the old, style of dredge, and the first that I shall mention is that of Messrs. Simons and Company, London Works, Renfrew, who have had very large experience in the construction of dredging apparatus. They have patented and constructed what they have called a hopper-dredge, combining in itself the advantages of a dredge for raising the material, and a screw hopper vessel for conveying it to the place of discharge, both of which services are performed by the same engines and the same crew. In contracted rivers and exposed bars and channels, the convenience of dispensing with hopper-punts lying alongside and chafing a large unwieldy dredge, must on some occasions be desirable, and the arrangement also secures the advantage of delivering the stuff at the level of a few feet above the deck, thus avoiding the long bucket-ladder and high level of discharge required for delivering into punts or barges, moored alongside the dredge. But the chief advantage which the patentees claim is a considerable saving of expense, not only in dredging, but in depositing the stuff at a distant point, such as Loch Long in the Clyde, and, if this saving can be satisfactorily established, there is much to commend the use of such craft as they have patented, especially in a crowded navigation.

Another of the recently suggested improvements is that by Mr. C. Randolph, who, in 1870, proposed that instead of the ordinary dredging-buckets, pipes should be lowered until they came into contact with the sand or mud at the bottom. The tops of these pipes were to be in communication with powerful centrifugal pumps, so that the velocity of the inflowing water through the pipes could be made so great as to carry with it a large percentage of the sand or mud from the bottom, and when the solid matter, and the water in which it is suspended, were raised to the desired height, they would flow freely to any required place for deposit of the suspended material. Mr. Thomas Stevenson, in his book on Harbours,[1] states that Mr. Duncan, the late engineer of the Clyde Navigation, had given him an extract from a report by M. Le Ferme, dated 30th September 1859, on the result of a silt pump of 20 horse-power, sunk 18 inches into the mud, proposed by M. Gache at St. Nazarie, which did its work effectively.

Another arrangement is that of raising the material by buckets in the ordinary way, and thereafter receiving it in a vessel and floating it off by pipes to the place of deposit. This of course can only be done where the place of deposit is close to the spot whence the material is dredged. Two plans have been proposed for effecting this. One of these has been used on the Amsterdam Canal, where the stuff is discharged from the buckets into a vertical cylinder, and is there mingled with water by a revolving Woodford-pump, and sent off under pressure to the place of deposit in a semi-fluid state. At the Amsterdam Canal this was done by pipes made of timber, and hooped with iron like barrels. These wooden cylinders were made in lengths of about 15 feet, connected with leather joints, and floated on the surface of the water, conveying the stuff to the requisite distance, like the hose of a fire engine, under a head of pressure, I believe, of 4 or 5 feet, and depositing it over the banks of the canal. A somewhat similar process was used on the Suez Canal,

Dredging at Amsterdam and Suez Canals.

---

[1] *The Construction of Harbours,* by T. Stevenson, Civil Engineer.

not, however, by using pumps, but simply by running the stuff to the banks on steeply inclined shoots, which were supplied with water when the material raised did not contain sufficient water to cause it to run freely. Another plan is that on which Messrs. Simons and Company are now constructing a dredge for Egypt, in which the stuff is to be propelled by jet-pumps, forcing a powerful stream of water through the discharge pipes. The dredgings will be conveyed by this means for a distance of two or three hundred feet on either side of the dredger, and at a considerable elevation.

It is obvious, however, that these arrangements can only be applied in situations where the material to be excavated is not of a hard nature, and where the place of deposit is close at hand. I can conceive, for example, in keeping clear the Suez Canal, that such appliances may be very useful, as the soft deposit of the canal has only to be raised and projected over the banks on either side. But this is not the place to discuss the claims of different inventors, which perhaps can only be settled by the actual performance of these arrangements when fully tested by practice. Having thus briefly noticed them I shall not dwell further on the subject, but conclude with a few practical observations on dredging as more immediately applicable to the rivers of this country.

Longitudinal and cross dredging.

In river dredging two systems are pursued. One plan consists in excavating a series of longitudinal furrows parallel to the axis of the stream, the other in dredging cross furrows from side to side of the river. It is found that inequalities are left between the longitudinal furrows, when that system is practised, which do not occur to the same extent in side or *cross dredging*, and I have invariably found cross dredging to leave the most uniform bottom. To explain the difference between the two systems of dredging it may be stated, that in either case the dredge is moored from the head and stern by chains about 250 fathoms in length. These chains, in improved dredges, are wound round windlasses worked by the engine, so that the vessel can be moved ahead or astern by simply throwing them into or out of gear. In longitudinal dredging the vessel is worked forward by the head chain while the buckets are at the same time performing the excavation; so that a longitudinal trench is made in the bottom of the river. When the dredge has proceeded a certain length, it is stopped and permitted to drop down and commence a new longitudinal furrow, parallel to the first one. In cross dredging, on the other hand, the vessel is supplied with two additional moorings, one on either side, and these chains are, like the head and stern chains, wound round barrels wrought by the engine. In commencing to work by cross dredging, we may suppose the vessel to be at one side of the channel to be excavated. The bucket-frame is set in motion, but instead of the dredge being drawn forward by the head chain, she is drawn to the opposite side of the river by the side chain, and having reached the extent of her work in that direction, she is then drawn a few feet forward by the head chain, and, the bucket-frame being still in motion, the vessel is hauled back again by the opposite chains to the side whence she started. By means of this transverse motion of the dredge a series of cross furrows is made; she takes out the whole excavation from side to side as she goes on, and leaves no protuberances such as are found to exist between the furrows of longitudinal dredging, even where it is executed with great care. The two systems will be best explained by reference to fig. 43, where A and B are the head and stern moorings, and D and C the side

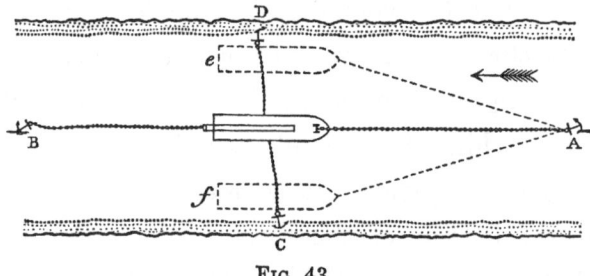

FIG. 43.

moorings; the arc ef represents the course of the vessel in cross dredging; while in longitudinal dredging, as already explained, she is drawn forward towards A, and again dropped down to commence a new longitudinal furrow.

In some cases, however, the bottom is found to be too hard to be dredged until it has been to some extent loosened and broken up. Thus at Newry, Mr. Rennie, after blasting the bottom in a depth of from 6 to 8 feet at low water, removed the material by dredging, at an expense of from 4s. to 5s. per cubic yard. The same process was adopted by Messrs. Stevenson at the bar of the Erne at Ballyshannon, where, in a situation exposed to a heavy sea, large quantities of boulder stones were blasted, and afterwards raised by a dredger worked by hand at a cost of about 10s. 6d. per cubic yard. But the most extensive application of blasting, preparatory to dredging, of which I am aware, was that on the works for improving the Severn, by Sir William Cubitt, of which an interesting and instructive account is given by Mr. George Edwards, in a paper addressed to the Institution of Civil Engineers, from which the following particulars are taken :[1]—

" It appears that a succession of marl beds, varying from 100 yards to half a mile in length, were found in the channel of the Severn, which proved too hard for being dredged, the whole quantity that could be raised being only 50 or 60 tons per day; while the machinery of the dredges employed was constantly giving way. Attempts were first made to drive iron rods into the marl bed, and to break it up; a second attempt was made to loosen it by dragging across its surface an instrument like a strong plough. But these plans proving unsuccessful, it was determined to blast the whole surface to be operated on. The marl was very dense, its weight being 146 lbs. per cubic foot ;[2] and it was determined to drill perpendicular bores, 6 feet apart, to the depth of 2 feet below the level of the bottom to be dredged out. The bores were made in the following manner, from floating rafts moored in the river :—Pipes of $\frac{3}{16}$-inch wrought-iron, $3\frac{1}{2}$ inches diameter, were driven a few inches into the marl. Through these pipes holes were bored, first with a $1\frac{1}{2}$-inch jumper, and then with an auger. The holes were bored 2 feet below

---

[1] " Account of Blasting on the Severn," by George Edwards, C.E. (*Minutes of Proceedings of Institution of Civil Engineers*, vol. iv. p. 361.)

[2] Clay weighs about 109 lb., and sandstone about 155 lb., per cubic foot.

the proposed bottom of the dredging, as it was expected
that each shot would dislocate or break in pieces a mass
of marl of a conical form, of which the bore-hole would be
the centre and its bottom the apex; so that the adjoining
shots would leave between them a pyramidal piece of
marl where the powder would have produced little or no

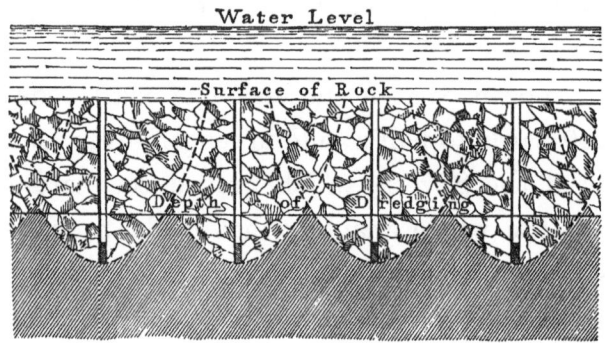

FIG. 44.

effect. By carrying the shot-holes lower than the intended dredging, the apex only of this pyramid was left

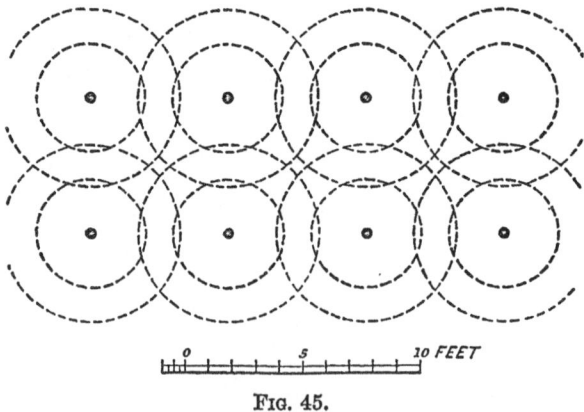

FIG. 45.

to be removed; and in practice this was found to form
but a small impediment. Fig. 44 is a section of the bore-

holes; and fig. 45 a plan, in which the inner dotted circles represent the diameters of the broken spaces at the level of the bottom of dredging. The cartridges were formed in the ordinary way, with canvas, and fired with Bickford's fuse. The weight of powder used for bore-holes of 4 feet, 4 feet 6 inches, and 5 feet, were respectively 2 lbs., 3 lbs., and 4 lbs. The effect of the shot was generally to lift the pipes, which were secured by ropes to the rafts, a few inches. Mr. Edwards says that not one in a hundred shots missed fire, and these shots were generally saved by the following singular expedient :—The pointed end of an iron bar, $\frac{5}{8}$-inch diameter, was made red-hot, and being put quickly through the water, and driven through the tamping as rapidly as possible, was, in nine cases out of ten, sufficiently hot to ignite the gunpowder and fire the shot.

" The cost of each shot is calculated as follows :—

	£	s.	d.
Use of material, ...	£0	1	0
Labour, ...	0	3	3
Pitched bag for charge, ...	0	0	3
3 lbs. of powder at $5\frac{1}{2}$d., ...	0	1	$4\frac{1}{2}$
15 feet of patent fuse at $\frac{6}{10}$ths of a penny, .	0	0	9
Pitch, tallow, twine, coals, etc., . . .	0	0	$4\frac{1}{2}$
Cost per shot, . .	£0	7	0

Each shot loosened and prepared for dredging about 4 cubic yards; so that the cost for blasting was 1s. 9d. per yard. The cost of dredging the material, after it had been thus prepared, was 2s. 3d.; making the whole charge for removing the marl 4s. per cubic yard."

Mr. J. Coode, in August 1871, made some interesting experiments on blasting below water, at St. Heliers,

Jersey, the result of which he has kindly communicated to me. The object Mr. Coode had in view was to demonstrate the feasibility of removing a mass of rock lying below low water by means of compressed gun-cotton, *without drilling holes for the insertion of the charges,* as in ordinary quarry blasting.

The rock upon which the trials were made at Jersey consisted of a hard syenite, with veins of trap; the rock, as a whole, was very compact, but numerous small "heads" or joints were interspersed throughout the mass —a circumstance, doubtless, greatly in favour of the rending or "shivering" action peculiar to gun-cotton in the compressed form.

Tin canisters, containing charges of 5 lbs. and 10 lbs. of compressed gun-cotton were prepared for these trials by Messrs. Prentice, of Stowmarket. They were fired simultaneously, in sets of three at a time, by means of a magneto-electric apparatus. The charges were placed upon the surface of the rock to be acted upon, and, as far as practicable, in crevices or hollows. They were fired at such times as to insure a great "head" of water, in order to obtain the greatest possible advantage from the downward action of the explosive compound. The results of the experiments at St. Heliers led Mr. Coode to believe that about 2 tons of the rock below low water at that place may be blasted and rendered in a fit state for lifting by each lb. of gun-cotton employed in the manner above described. It was, and still is, Mr. Coode's intention to try the comparative effect of "litho-fracteur" and gun-cotton, but he has hitherto been prevented from doing so by the stringency of the "Nitro-Glycerine Act" of 1869, which makes it a penal offence to import, convey, or manufacture nitro-glycerine or any of its compounds within the United Kingdom, unless by special license from the Secretary of State, but no such license can be at present obtained as regards litho-fracteur.

In some cases dredging has to be conducted in exposed situations, such as the deepening of the "flats" at Londonderry and the bar at Carlingford. The process of dredging at the Foyle could not be conducted when the waves exceeded $2\frac{1}{2}$ feet; and Mr. Barton at Dundalk so far confirms this, as he estimates a swell of 2 feet as the highest to work in. Mr. Barton states that the bar at Carlingford, which is very exposed, consists of hard blue clay, with a coating of large boulders. The dredger employed was built by Messrs. Simons and Co. of Renfrew. She is 157 feet in length, 27 feet beam, and 9 feet 6 inches in depth. Her bucket-ladder is 90 feet long, with 42 buckets, and she dredges in 35 feet water. She has an engine of 50 horse power, and her work varies from 100 to 4000 tons per day, but her average work is 1000 tons per day, and she has brought up with her buckets boulders weighing 3 tons. All stones above that weight are removed by divers.

Dredging in exposed situations.

6. *Excavation.*

But there are cases where the bottom cannot be dredged, and where it is necessary to have recourse to other appliances for its removal, such as the diving-bell

Diving-bell. or diving-helmet, and cofferdams. The diving-bell has, in conjunction with dredging, been much used on the Clyde, and Mr. Bald gives the following account of the operation as conducted on that river :—

"Between Erskine Ferry and the New Shot Isle the bed of the Clyde, for a distance of 2000 yards, was greatly encumbered with boulders, which were highly injurious to vessels if they grounded there; and frequently large ships, in being tugged through this part of the river-channel, had their copper bottoms injured when they touched the rocky channel-bed. In deepening and clearing this part of the river, two diving-bells were employed, and one, and sometimes two, steam dredgers. The clearing and deepening of this channel was exceedingly severe on the machinery and working gear of the steam dredgers; the speed of the engines was therefore governed by the nature of the material in the bottom, and although the iron-work frequently gave way, yet spare links and buckets being always ready to replace those which broke, there was little interruption to the continuous working of the dredgers. When the dredgers had cleared away the material which covered the boulders in the bottom of the channel, the diving-bell boats were worked over the ground so cleared, removing all the larger boulders; and when that part of the channel had been cleared of them, the dredgers went again over the same bottom, removing all the lighter material from the heads of the lower boulders, preparatory to the bells commencing again; and these operations were continued until the necessary depth was attained.

" The buckets of the steam dredgers, in working along the bottom, always slipped over the head of the large boulders, which the diving-bells alone could lift and remove. Some of these masses of trap or whinstone were 4 and 5 tons in weight, and from their rounded forms and smooth surfaces, it was evident that they had been brought from some distance. Some of them were of sandstone, but they were more angular than the trap boulders. Quantities of these boulders, lifted from the bed of the channel, might be seen lying along the sides of the river; and many of them had since been split and broken up by gunpowder for repairing the river dykes. The tops of some of the large boulders lifted from the bed of the channel were found grooved to a depth of about an inch or more, by the ships' keels having been rubbing over them; and metallic particles were distinctly to be seen upon their surface. In removing these boulders from the bed of the channel, the diving-bell men found numerous fragments of copper and iron which had been torn off the ships' bottoms and keels by the large stones; but latterly this had not been the case, as great progress had been made in the removing of the boulders, and the deepening of the channel."

Large isolated masses of stone have also been removed By Floatation *en masse* by fixing louises in them, and raising them by floatation. On the Tay many boulders were raised in this manner, one of which weighed upwards of 50 tons. Where a large area and considerable depth of solid rock has to be removed, it will generally be found most advantageous to employ cofferdams; but the chief objections to the use

of dams in the narrow channels of rivers are the inconvenience they cause to shipping by increasing the currents, so as to render navigation past them difficult or
dangerous, and the obstruction they offer to the discharge
of water when the river is in high flood.   An illustration
of this occurred at the Ribble, near Preston, where a
solid band of red sandstone, upwards of 300 yards in
length, crossed the river and restricted the navigation,
even at highest springs, to "lighters" or "flats" of
small draught ; and, in order to gain the requisite depth,
it was found that an excavation of the maximum depth
of 13 feet 6 inches must be made through this solid bed
of rock.   It was originally intended to form a temporary
channel, and to divert the river while the excavation was
being made, but the rock was found to extend beyond
the river's bank, and even to rise in level on the adjoining ground, and there seemed to be no other course open
Cofferdams. but to erect a cofferdam in the navigable channel.   In
doing this, however, there were two difficulties to contend
with ; for not only had the dam to be fixed on a rocky
bottom, but the narrowness of the river, and the necessity
of preserving a channel for flood-water, and occasional
passage of a "lighter," left only a very narrow space for
a foundation on which to construct it.   To overcome this
double object, I designed a cofferdam, which was found
to answer the requirements of the case.[1]   It consisted, as
shown in fig. 46, of two rows of iron rods, 3 feet apart,
jumped into the rocky bottom, and supporting two

[1] "Description of a Cofferdam adapted to a Hard Bottom," by David Stevenson, C.E. (*Trans. of Inst. of Civil Engineers*, vol. iii. p. 377.)

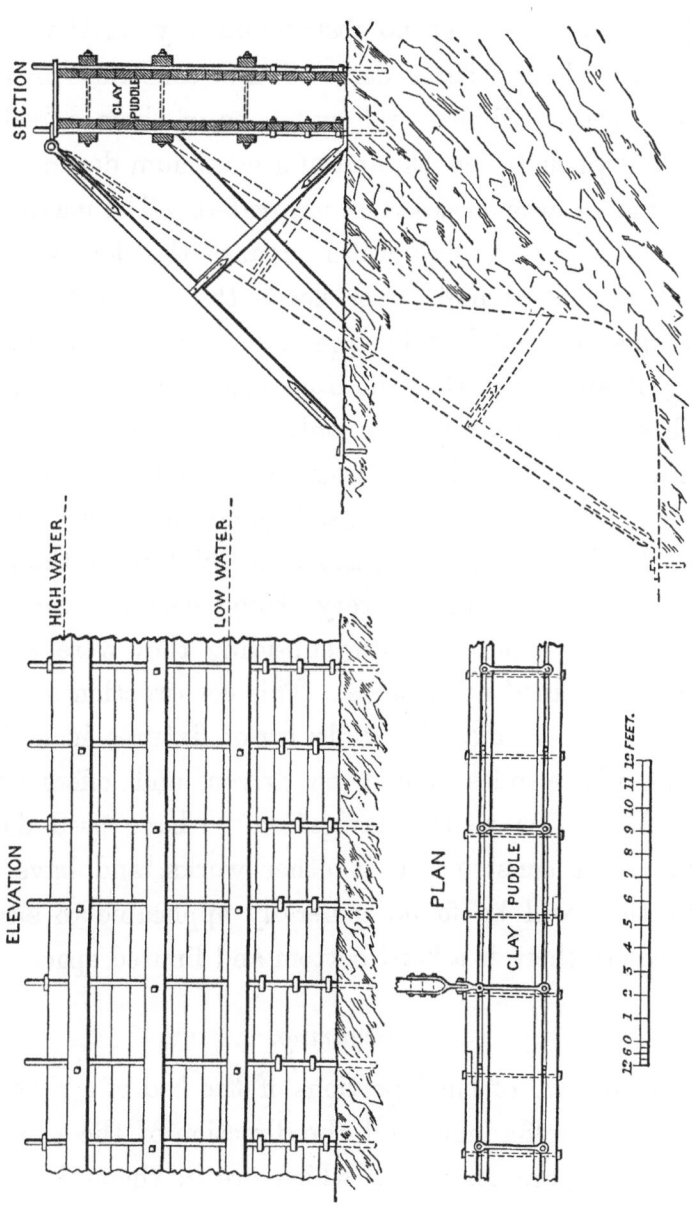

SECTION

CLAY PUDDLE

HIGH WATER

LOW WATER

ELEVATION

PLAN

CLAY PUDDLE

12 6 0 1 2 3 4 5 6 7 8 9 10 11 12 FEET.

FIG. 46.—COFFERDAM FOR A HARD BOTTOM.

linings of planking, the intermediate space being filled with clay, and the whole structure being stayed from the inside, so as to present no obstruction beyond the outer line of the dam. Three dams of this construction were formed in the Ribble; and by means of them a bed of rock, 300 yards in length, and of a maximum depth of 13 feet 6 inches, was successfully excavated. The maximum depth of water at high water against the dam was 16 feet, but in very high river floods the whole dam was sometimes completely submerged; but on the water subsiding, it was found that the iron rods, on which alone its stability depended, although only jumped 15 inches into the rock, were not drawn from their fixtures. On one occasion, on visiting the works, I found the river in high flood, the dams submerged, not even the tops of the iron rods being visible, and a very strong current sweeping over them; but on the water subsiding they were found to have sustained no damage. This construction of dam completely overcomes the difficulty of fixtures in a hard bottom, where piles cannot be driven, and offers very little obstruction to the navigation. I have used dams of the same construction in other works, and have no doubt they will be found generally applicable to situations where there is a hard bottom and limited space.

7. *Scouring.*

The removal of hard portions of the bed of a river by dredging or cofferdams, and the direction of the channel by low walls, are operations which are in themselves improvements; but they further operate beneficially in causing the currents to scour the softer parts of the river's bed. Thus I have found that by dredging a few hundred yards of hard material, or erecting a short wall, thousands of tons of soft material are scoured away by the action of the current alone. In all river improvements this is an effect which should be fully taken into consideration by the engineer, especially in forming estimates; and its importance will be apparent on inspecting the section of the river Lune (Plate VIII.) By dredging the upper shoals of that river, which are marked in hatched lines in the section, the whole lower part of the river was deepened by the natural scour, without entailing any expense in its removal. To facilitate this scour, a species of harrow has sometimes been applied, which is drawn to and fro by a tug-steamer across the bank to be removed. This system was extensively employed by Captain Denham in opening the Victoria Channel at the Mersey; it was also employed by Messrs. Stevenson at the Tay; but it is obvious that it can only be advantageously used where there is deep water in the immediate neighbourhood of the bank to be removed, in which the sand and mud disturbed by the harrow, and carried off by the current, may be deposited. I have found that the process of natural scouring has, in some situations, continued in operation for many years after the completion of the original work, the low-water level of the river continuing gradually to sink; and, as this process goes on, it sometimes happens that hard portions of the bottom, originally covered, become gradually exposed. Such obstructions are, in fact, hard portions of the bed brought to light, in consequence of the improvement of the river, and must not be mistaken for accumulations due to ill-regulated currents. It is necessary, however, that such hard portions should be removed as soon as they appear, otherwise they disturb the currents and occasion shoals. Whenever the depth due to the currents acting in their improved direction has been reached such obstructions will cease to present themselves.

8. *Reducing the Inclination of the Bed.*

The existence of a not immoderate amount of fall or slope on the low-water line of a river may always be regarded by the engineer as affording good encouragement for its improvement. The slopes of rivers vary from a few inches to several feet per mile, as will be seen from the tabular list appended to this book, of the physical characteristics of rivers in which the inclinations of several rivers are given. Du Buat considers 1 in 500,000 to be the smallest possible rate of inclination that can be given to a canal to produce sensible motion. In dealing with rivers I should say, from my own experience, that the engineer may calculate on *reducing the slopes of tidal navigations to about 3 or 4 inches per mile,* which is equal to $\frac{1}{15840}$, and that *they should not, if possible, exceed 10 inches per mile,* which is equal to a gradient of $\frac{1}{6336}$.* The lowering of the low-water line, and consequent flattening of the slope or inclination, acts beneficially both on the *tidal propagation* and the *scour.* As regards tidal phenomena it will be found that in all rivers whose

* The slopes of " the rapids " immediately above the Falls of Niagara are said to be 1 in 52·8.

beds and low-water lines have been lowered, the rate of tidal propagation has been increased, and the duration of the tide in the river has been prolonged, to the benefit of navigation, as will be explained hereafter, when we treat of rivers that have been improved. It will also be found that the scouring power has in some cases been enormously increased, and made to act in the most beneficial way for the channel, a result which in river engineering can hardly be over-estimated. In order to illustrate this it is only necessary to point out that the mere cubic contents dredged from a ford or shoal form no measure of the gain of tidal water due to the operation, as explained under " Scouring," because the removal of such an obstruction has the effect of lowering the low-water line for a certain distance on either side of it, and the extent of the lowering will obviously depend on the original amount of slope ; if very steep the extent will be small, if gentle it will be greater. But in either case, whether small or great, the whole of the wedge-sloped volumes included between the old and new water surfaces, both above and below the obstacle that has been removed, give a clear gain of tide water, and the cubic contents of these spaces greatly exceed the cubic quantity of material removed from the ford by dredging.

Proceeding on actual calculations of comparative sections of the river Tay, before and after the execution of the works, I found that by excavating about 500,000 cubic yards of gravel the low-water line was lowered so much as to admit an additional quantity of *sea*-water,

P

amounting on an average to not less than one million cubic yards to be propelled into and again withdrawn from that part of the river which lies above Newburgh, *during every tide.* This quantity is equal to nearly two hours' discharge of the Tay in its ordinary state, and it, therefore, follows that one way at least, of representing the amount of increase during a year is to compare it to two months' constant ordinary flow of the river.

Velocities of tide-currents.

The velocity of the stream at low water depends on the slope, and in our navigable rivers it rarely exceeds 3 or 4 miles per hour, and of course is considerably higher when the river is in flood. The tide-currents, however, attain a higher velocity than the ordinary flow of the river. But I have found in almost every case I have had to investigate that the rate of the tide-current was greatly exaggerated. A current of 6 or 7 knots an hour in the fairway is really hardly navigable. Even in the Dee, where the rise of tide is great and the currents are very rapid, I do not think they much exceed 5 miles an hour, and in the Tay above Newburgh, and rivers of that class, where the rise of tide is not so great, I do not think the current exceeds 4 miles. At the Severn, again, where there is a rise of 40 feet of tide, the current is said by Captain Beechey to reach 9 miles per hour; at the Mersey, with a very high tide, it was found by Captain Denham to run 7 miles; and Captain Otter, in his survey of the Pentland Firth, gives the velocity off the Pentland Skerries at 10·6 nautical miles per hour, which I believe to be the highest tide-current ever observed.

The following may be taken as the surface velocities of the currents in different rivers :—

VELOCITIES OF THE CURRENTS IN DIFFERENT RIVERS.

Name.	Per Hour.		Authority.
	Miles.	yds.	
Mississippi, ...	5	0	C. Ellet.
Clyde, between Glasgow and junction of Cart, during ebb,	0	1576	W. Bald.
Do., flood, ...	0	771	Do.
Do., from junction of Cart to Dumbarton, ebb, . .	1	1069	Do.
Do., flood, ...	0	1561	Do.
Do., during high floods below Glasgow harbour, ebb, . .	2	1613	Do.
Do. at narrow places during floods, ...	3	1148	Do.
	Miles.		
Wear, spring-tide, ebb, . .	$1\frac{1}{2}$ to $2\frac{1}{2}$		J. Murray, C.E.
Do., neap-tides, „ . .	1 to $1\frac{3}{4}$		Do.
Do., flood-tides, . . .	1 to 2		Do.
	Knots.		
Tay at Buddonness, spring-tides,	2 to $2\frac{1}{2}$		North Sea Pilot.
	Miles.		
Do. at Perth, ...	3·09		Messrs. Stevenson.
Willowgate at Perth, . .	1·55		Do.
Dornoch Firth, Meikleferry, flood, ...	2·63		Do.
Do. do. ebb,	2·55		Do.
Tay at Mugdrum, flood and ebb,	2 to $2\frac{1}{2}$		Do.
Thames, ...	2 to $2\frac{3}{4}$		G. Rennie.

The beneficial effect of the works I have described Effect of works may be summarized as follows :—

First, To depress the level of the low-water line.

Second, To increase the range of tide.

Third, To accelerate the propagation of the tide through the channel of the river.

Fourth, To prolong the duration of the tide in the river.

Fifth, To equalize the velocity of the tidal currents, removing rapids and bores.

Sixth, To add to the beneficial scouring power of the river ; and

Seventh, To increase the navigable depth.

CHAPTER IX.

APPLICATION OF THESE WORKS IN PRACTICE.

On the River Tay : description of works executed ; alterations produced on the
slopes, and rates of propagation and duration of tidal influence—The River
Forth : description of works ; their effect on the propagation of the tide—
Laws of tidal propagation in rivers generally—The River Ribble : works executed and their effects—The River Lune : works executed and their effects—
The River Clyde : its former and present condition—The River Tees : works
executed and their effects.

I PROPOSE now, by reference to examples in actual
practice, to show the beneficial effect produced by the
works specified in the last chapter, in some navigations
where they have been adopted, and the first example to
which I shall allude is the river Tay. The original
design for its improvement was made by Messrs. Robert
and Alan Stevenson, and the works were commenced in
1833, and afterwards carried out partly under my direction ; and I know of no instance where the improvements,
effected by particular works, are more fully and satisfactorily demonstrated, by a comparison of observations
made *previously* and *subsequently* to their execution. A
notice of them will, I think, be interesting to engineers,
not so much as a record of what was done on the Tay,
but as affording an illustration of the relation that exists

between the forms of a river's bed and its tidal phenomena, for it will be clearly seen in the case of the Tay, that in altering its bed and the inclination of its surface, many marked changes were effected on its tidal phenomena, while in those parts of the river where no works were executed the tidal phenomena were not altered. I propose, therefore, to enter into some detail as to the obstructions met with on the Tay, the means employed for their removal, and the effects produced on the tidal currents, as illustrating the subject of river improvements generally.

RIVER TAY.

The river Tay, with its numerous tributaries, receives the drainage water of a district of Scotland amounting to 2283 square miles, as measured on Arrowsmith's map. Its *mean* discharge has been ascertained to be 274.000 cubic feet, or 7645 tons of water per minute. It is navigable as far as Perth, which is 22 miles from Dundee, and 32 from the German Ocean. The different points on the river hereafter to be referred to will be seen in the chart given in Plate XIV.

Before the commencement of the works, certain ridges, called "fords," stretched across the bed of the river at different points between Perth and Newburgh, and obstructed the passage to such a degree that vessels drawing from 10 to 11 feet could not, during the highest tides, make their way up to Perth without great difficulty. The depth of water on these fords, the most objectionable of which were six in number, varied from 1 foot 9 inches to 2 feet 6 inches at low, and 11 feet 9 inches to 14 feet at high water of spring-tides; so that the regulating navigable depth, under the most favourable circumstances, could not be reckoned at more than 11 feet. In addition to the shallowness of the water, many detached boulders lay scattered over the bottom. Numerous "fishing cairns," or collections of stones and gravel, had also been laid down, without regard to any object but the special one in which the salmon-fishers were interested, and in many cases they formed very prominent and dangerous obstructions to vessels. The chief disadvantage experienced by vessels in the unimproved state of the river was the risk of their being detained by grounding, or being otherwise obstructed at these defective places, so as to lose the tide at Perth,—a misfortune which, at times when the tides were falling from springs to neaps, often led to the necessity either of lightening the vessel, or of detaining her till the succeeding springs afforded sufficient depth for passing the fords. The great object aimed at, therefore, was to remove every cause of detention, and facilitate the propagation of the tidal wave in the upper part of the river, so that inward-bound vessels might take the first of the flood to enable them to reach Perth in one tide. Nor was it, indeed, less important to remove every obstacle that might prevent outward-bound vessels from reaching Newburgh, and the more open and deep parts of the navigation, before low-water of the tide with which they left Perth.

The works undertaken by the Harbour Commissioners

of Perth for the purpose of remedying the evils alluded to, and which extended over six working seasons, may be briefly described as follows :—

1*st*, The fords, and many intermediate shallows, were deepened by steam dredging; and the system of harrowing was employed in some of the softer banks in the lower part of the river. The large detached boulders and "fishing cairns," which obstructed the passage of vessels, were also removed.

2*d*, Three subsidiary channels, or offshoots from the main stream, at Sleepless, Darry, and Balhepburn islands, the positions of which will be seen on the plan, were shut up by embankments formed of the produce of the dredging, so as to confine the whole of the water to the navigable channel, and the banks of the navigable channel were widened to receive the additional quantity of water which they had to discharge.

3*d*, In some places the banks on either side of the river beyond low-water mark, where much contracted, were excavated, in order to equalize the currents, by allowing sufficient space for the free passage of the water; and this was more especially done on the shores opposite Sleepless and Darry islands, where the shutting up of the secondary channels rendered it more necessary.

The benefit to the navigation in consequence of the completion of these works has been of a twofold kind; for not only has the depth of water been materially increased by actual deepening of the water-way, and the removal of numerous obstructions from the bed of the river, but a clearer and a freer passage has been made for the flow of the tide, which now begins to rise at Perth much sooner than before; and as the time of high water is unaltered, the advantages of increased depth due to the presence of the tide is proportionally increased throughout the whole range of the navigation; or, in other words, the *duration of tidal influence has been prolonged.*

The depths at the shallowest places were pretty nearly equalized, being 5 feet at low and 15 feet at high water, of ordinary spring-tides, instead, as formerly, of 1 foot 9 inches at low and 11 feet at high water. Steamers of small draught of water can now therefore ply at *low water*, and vessels drawing 14 feet can now come up to Perth in *one tide* with ease and safety.

Such was the state of matters in 1845, when, during what has been called the *railway mania* of that period, two companies proposed to cross the Tay by bridges between Newburgh and Perth. These schemes were naturally regarded by the navigation authorities as a great aggression on the rights of the public as proprietors of the highway of the river. In preparing a report in support of the views of the conservators of the river, it occurred to me that it was hardly sufficient to aver that so much gravel had been dredged, and so many fords had been removed, but that, if not essential, it would at least be interesting to know what effect the works had produced on the tidal action of the river. I accordingly made an elaborate analysis of the tidal observations, extending over ten years, which not only showed substantial improvements in the state of the river, but gave highly valuable information, which may

be held to be of general application. I am not aware
that, previous to the publication of the Tay observations,
in the report to the Admiralty, on the railway bridge
to which I have alluded, there had been any statement
demonstrating the alterations on tidal flow produced
by removing obstacles to its propagation; I accordingly
submitted the result to the Royal Society of Edinburgh;[1]
and as the subject is generally interesting in connexion
with river engineering, I make no apology for giving the
details in this treatise.

The tidal observations to which I have referred were
made at various times during a period of ten years, from
1833 to 1844 inclusive, throughout the river and firth
of Tay, at the following stations, viz., Dundee, which is
marked No. 1 on the plan, Plate XIV.; Balmerino, No.
2; Flisk Point, No. 3; Balmbreich Castle, No. 4; Newburgh, No. 5; Carpow, No. 6; Kinfauns, No. 7; and
Perth tide harbour, No. 8. The general results deduced from these observations are given in the following
tables, and show, by the favourable change which has
been effected in the tidal phenomena of the estuary, that
the works executed fully answered the intended end.

1. *Propagation of Tidal Wave.*

The following table of elapsed times between arrival
of the tide-wave, or commencement of the tidal flow, at
the following stations, during *spring* tides in 1833 and
1834, shows the rate of its propagation:—

[1] *Proceedings of the Royal Society of Edinburgh* for 1845.

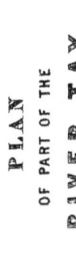

PLATE XIV.

PLAN

OF PART OF THE

RIVER TAY.

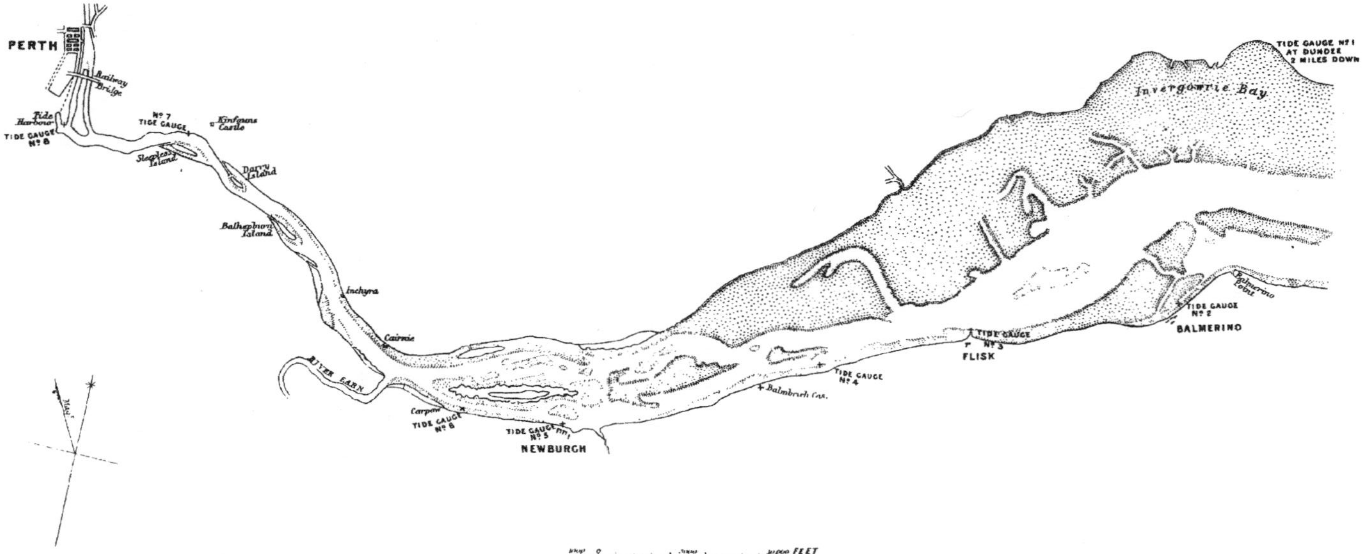

Invergowrie Bay

TIDE GAUGE Nº 1
AT DUNDEE
AT MILES DOWN

TIDE GAUGE
Nº 2

SALMERINO

TIDE GAUGE
Nº 3

FLISK

TIDE GAUGE
Nº 4

Ballenbreich Cas.

NEWBURGH

TIDE GAUGE Nº 5

TIDE GAUGE
Nº 6

Carpow

RIVER EARN

Inchyra

Mugdrum Island

Newburgh House

Elcho Castle

Seggieden

Kinfauns Castle

Nº 7
TIDE GAUGE

Railway Bridge

PERTH

TIDE GAUGE
Nº 8

scale FEET

Published by A. & C. Black, Edinburgh.

	Time. H. M.	Distance in Miles.	Rate of Tide-Wave in Miles per Hour.
Dundee to Balmerino, . .	0 16	5·00	18·75
Balmerino to Flisk Point, . .	0 29	2·93	6·06
Flisk Point to Balmbreich, . .	0 26	2·04	4·69
Balmbreich to Newburgh, . .	0 53	3·42	3·86
Newburgh to Perth (tide harbour),	2 30	8·56	3·42

The result of observations made in 1842, 1843, and 1844, on spring-tides, give the same velocity, as above stated, between Dundee and Newburgh, where no works had been done, and the following rates between Newburgh and Perth, below which place works had been executed :—

	Time. H. M.	Distance in Miles.	Rate of Tide-Wave in Miles per Hour.
Newburgh to Carpow, . .	0 25	1·33	3·17
Carpow to Kinfauns, . .	0 55	4·92	5·36
Kinfauns to Perth (tide harbour),	0 20	2·32	6·93
Giving as a mean for the whole distance from Newburgh to Perth in 1844, . . .	1 40	8·56	5·13
Time from Newburgh to Perth in 1833, ...	2 30	8·56	3·42

Thus showing an increase in the velocity of the tide-wave in the upper part of the river, which was improved, of more than 1⅔ mile per hour.

The difference of the time in *neap* tides between Newburgh and Perth, in 1844, was 1 h. 53 m.

2. *High-Water Level.*

The levels of the surface of high water at different stations throughout the river have been found to be unchanged, and the following results refer to the years 1833 and 1844 :—

From Flisk Point to Balmbreich there is a fall of　5　in.,⎫
　„　Balmbreich to Newburgh there is a rise of　7½　„　⎬ spring-tides.
　„　Newburgh to Perth (tide harbour) there is⎪
　　　a rise of　.　.　.　.　.　.18　„　⎭

From Flisk to Balmbreich there is a fall of　.　2½　„　⎫
　„　Balmbreich to Newburgh there is a rise of　6　„　⎬ neap-tides.
　„　Newburgh to Perth (tide harbour) there is⎪
　　　a rise of　.　.　.　.　.　.12　„　⎭

3. *Low-water Level.*

Rise on the Surface of Low-water (Spring-Tides) in 1833.

	Ft. In.	Distance in Miles.	Rate of Slope per Mile in Inches.	Rate of Tide in Miles per Hour.
Flisk to Balmbreich there was a rise of . . .	0 4	2·04	1·95	4·69
Balmbreich to Newburgh, a rise of	2 8	3·42	9·35	3·86
Newburgh to Perth (tide harbour), a rise of . .	4 0	8·56	5·06	3·42

Rise on the Low-water of Spring-Tides in 1844.

	Ft. In.	Distance in Miles.	Rate of Slope per Mile in Inches.	Rate of Tide in Miles per Hour.
Newburgh to Carpow, there is a rise of . . .	0 5	1·33	3·75	3·17
Carpow to Perth, there is a rise of	1 7	7·23	2·63	...
Hence from Newburgh to Perth, 1844, the rise is . .	2 0	8·56	2·80	5·13

The result of the observations of 1844 thus gives a depression on the level of the low-water mark of 2 feet at Perth tide harbour, the level of low water at Newburgh being unaltered.

4. *Duration of Flood and Ebb.*

The results of observations in 1833 and 1844 at New-burgh, where no works were executed, showed that the

durations of flood and ebb tides at that place are unchanged. The times are as follows, being almost identical in both years :—

		H.	M.
Spring-tides flow, , ...		4	20
„ „ ebb, ...		7	20
Neap-tides flow, ...		4	30
„ „ ebb, ...		6	45

At Perth, in 1833 :—

Spring-tides flowed, ...		2	20
„ „ ebbed, ...		7	0
Neap-tides flowed, ...		3	15
„ „ ebbed, ...		7	0

At Perth, in 1844 :—

Spring-tides flowed, ...		3	10
„ „ ebbed, ...		7	0
Neap-tides flowed, ...		3	10
„ „ ebbed, ...		7	0
Increase in duration of flood in springs at Perth,		0	50

It will be observed from these tables that important
changes have taken place in the part of the river that has
been improved :—

First, The fall on the surface of the river from the
tide basin at Perth to Newburgh in the year 1833 was
4 *feet*, but after the works were executed it was only
2 *feet*.

Second, In 1833 the passage of the tidal wave from
Newburgh to Perth (8·56 miles) occupied 2 hours 30
minutes, being at the rate of 3·42 miles per hour ; but it
is now propagated between the same places in 1 hour 40
minutes, being at the rate of 5·13 miles per hour,—giving

a *decrease* in the time of 50 minutes, and an *increase* in the speed of the first wave of flood of more than 1⅔ mile per hour, since the commencement of the works.

Third, The spring-tides in 1833 at Perth flowed 2 hours 20 minutes, and ebbed 7 hours; but now the tide flows 3 hours 10 minutes, and ebbs 7 hours,—being an *increase* in the duration of flood of 50 minutes.

Fourth, It will further be noticed that on the part of the river between Dundee and Newburgh, where no works had been executed, the tidal phenomena remain unaltered.

RIVER FORTH.

The works on the Forth, executed under the direction of Messrs. D. and T. Stevenson, produced changes on the tidal phenomena, which, in connexion with those described on the Tay, are interesting and instructive as regards the propagation of the tide, and therefore I shall briefly allude to them. The river between Stirling and Alloa is very circuitous, the distance by the navigation being 10½ miles, while in the direct line it measures only 5 miles. The navigation was found to be impeded by seven fords or shallows which occur between Alloa and Stirling, and are composed of boulders, varying from a few pounds to several tons in weight, embedded in clay.

It was determined, in the first instance, to remove two of these obstructions, viz., the "Town" and the "Abbey" fords, which lie nearest to Stirling, and having the smallest depth of water, formed the greatest obstruction to the free

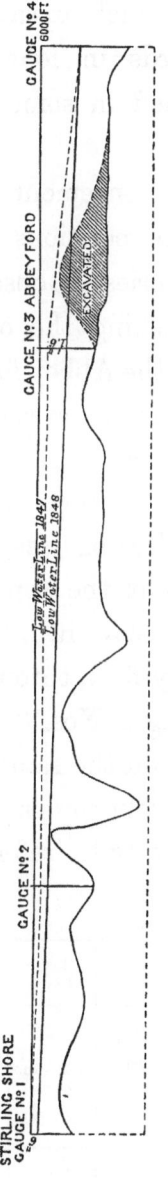

FIG. 47.—SECTION OF RIVER FORTH.

passage of vessels. The works were commenced at the lower end of the Abbey ford, and were carried regularly upwards. The new channel excavated through this ford was about 500 yards in length, and 75 feet in breadth, and was deepened in some places about 3 feet 6 inches.

Previous to the commencement of the work, tide-gauges were erected in the positions marked 1, 2, 3, and 4, in fig. 47, on which a series of observations was made for the purpose of establishing the original tidal pheno-mena of the river. After the Abbey ford was cut through, further observations were made on the same gauges; and it is to a comparison of these two sets of observations that I desire specially to refer. It is necessary to explain that gauge No. 1 is at Stirling quay, No. 2 about 500 yards farther down, No. 3 at the top of the Abbey ford, and No. 4 immediately below it. It will therefore be understood that the Abbey ford, through which a channel was cut, lies between gauges Nos. 3 and 4. The whole of the gauges were placed on the same level, so that their *readings* might be more easily compared; and the follow-ing are the results obtained with reference to the level of the low-water line :—

Levels of Low-water Line.	Gauge No. 4.		Gauge No. 3.		Gauge No. 2.		Gauge No. 1.	
	Ft.	in.	Ft.	in.	Ft.	in.	Ft.	in.
In 1847 the low-water line was found to stand at the following levels, ...	2	0	5	0	5	3	5	6
In 1840, ...	2	0	3	6	4	6	5	0
Depression, . .	0	0	1	6	0	9	0	6

From this tabular statement we find that the low-water line at No. 4, which is below the site of the works, remains unaltered, but that it has fallen 1 foot 6 inches at the top of the Abbey Ford (through which the cut has been made). It further appears that the formation of this cut has drained off the water, and lowered the surface 9 inches at gauge No. 2, and 6 inches at gauge No. 1, which is at Stirling. The former and present low-water lines and bed of the river are represented in fig. 47, in which is also shown by hatched lines the amount of excavation on the Abbey Ford. This general depression of the river has of course altered the slopes or inclinations formed by the surface of low water; the slope between 4 and 3 being decreased, while the inclinations between 3 and 2, and between 2 and 1, have been increased in the following ratios :—

Inclinations.	Distance. Feet.	1847. Inches per Mile.	1848. Inches per Mile.	Difference in 1848.
Inclination between 4 and 3, .	1550	122·5	61·3	− 61·2
Do. do. 3 and 2, .	3050	5·19	20·77	+15·58
Do. do. 2 and 1, .	1400	11·31	22·62	+11·31

Again, these changes on the low-water line have produced corresponding alterations on the velocities of the first wave of flood, which are found to be as follows :—

Velocities.	1847.	1848.	Difference.
	Minutes.	Minutes.	Minutes.
Time occupied by first wave of tide in passing between gauges Nos. 4 and 3, ...	24	8	− 16
Do. do. Nos. 3 and 2,	6	$11\frac{1}{2}$	$+5\frac{1}{2}$
Do. do. Nos. 2 and 1,	6	$8\frac{1}{2}$	$+2\frac{1}{2}$
Do. do. Nos. 4 and 1,	36	28	− 8

From this it appears that between Nos. 4 and 3 there is an acceleration of 16 minutes, while between 3 and 1 there is a retardation of 8 minutes, leaving the difference, or 8 minutes, as the actual amount of acceleration at Stirling, due to the removal of the ford and the lowering of the low-water level 6 inches at that place. The rates of propagation in miles per hour are as follows :—

Rates of Propagation.	1847.	1848.	Difference.
	Miles per Hour.	Miles per Hour.	
Rates of Propagation between Nos. 4 and 3, ...	·65	2·2	+ 1·55
Do. do. Nos. 3 and 2,	5·77	3·0	− 2·77
Do. do. Nos. 2 and 1,	2·65	1·87	− 0·78

Relations of the slopes and rates of tidal propagation.

These observations and results throw some additional light on the circumstances which modify the propagation of the tidal wave. The table of the results obtained at the Tay shows that the *decreased* inclination of the low-water lines of that river was attended by an *acceleration* of the velocity of the tidal wave ; and the above observations further show that a *retardation* has attended an *in-creased* inclination of the low-water line of the upper part of the Forth. From the foregoing tabular statements it will be seen that between gauges 4 and 3, where the slope has been decreased, the propagation has been accelerated ; while between 3 and 2, where, from the state of the works when the observations were made, it is found to have been increased, the rate of propagation had been sensibly retarded. It is worthy of remark, however, that the rates of propagation do not, either at the Tay or Forth, *bear any constant relation to the slopes*, but are modified by other circumstances ; in proof of which it will be found that the rate of propagation at the Forth between gauges 4 and 3, where the slope is 61·3 inches per mile, is actually greater than between gauges 2 and 1, where it is only 22·62 inches per mile. The circumstances of the Forth at this particular place are somewhat peculiar. Before the Abbey Ford was cut through, it acted as a dam extending across the river, and had the effect of increasing the depth at low water all the way up to Stirling. By cutting the channel through the ford, however, not only has the water been drained off and rendered shallow, but its surface has been broken by the projection of boulders from the bottom, which formerly were entirely covered ; and while this effect has taken place in the upper part of the river, a comparatively smooth cut, with regular sides and bottom, has been formed in the Abbey Ford, through which the river flows at low water in a body of considerable depth. I therefore attribute the slow propagation of the tide between

2 and 1 to the shallowness of the water and the very
rugged state of the bottom, which is in many places
completely studded with boulders, rising some above the
surface at low water, and others to within a few inches
of it; while the high velocity up the steep slope of the
ford is to be attributed—1st, To the depth of water
caused by the whole river being made to pass through a
comparatively narrow channel; 2d, To the rectangular
cross section of the cut; and 3d, To the smoothness of
the sides and bottom. At the Firth of Dornoch again, as
already noticed, between the Quarry and Bonar Bridge, a
distance of 1 mile, although the water is shallow and the
bottom rough, it is not, on the whole, more so than between gauges 1 and 2 on the Forth; but at the Dornoch
the slope on that mile is no less than 6 feet 6 inches, and
the rate of propagation is only two-thirds of a mile per
hour. Moreover, it was found that the tide did not begin
to show at Bonar until it had risen 6 feet 6 inches on the
gauge at the Quarry, being the exact difference of level
between the two points of observation.

Laws of tidal propagation in rivers.
These various results as to slopes and rates of propagation, as well as others which have come under my
notice, seem to justify the following deductions as to the
propagation of the tide-wave in rivers with sloping surfaces and irregular bottoms, which, as stated at page 158,
may be regarded as an addition to the laws formerly
stated as regulating tidal propagation. These results are
as follows :—

1st, That a decrease of slope is followed by an acceleration of the rate of propagation of the tidal wave.

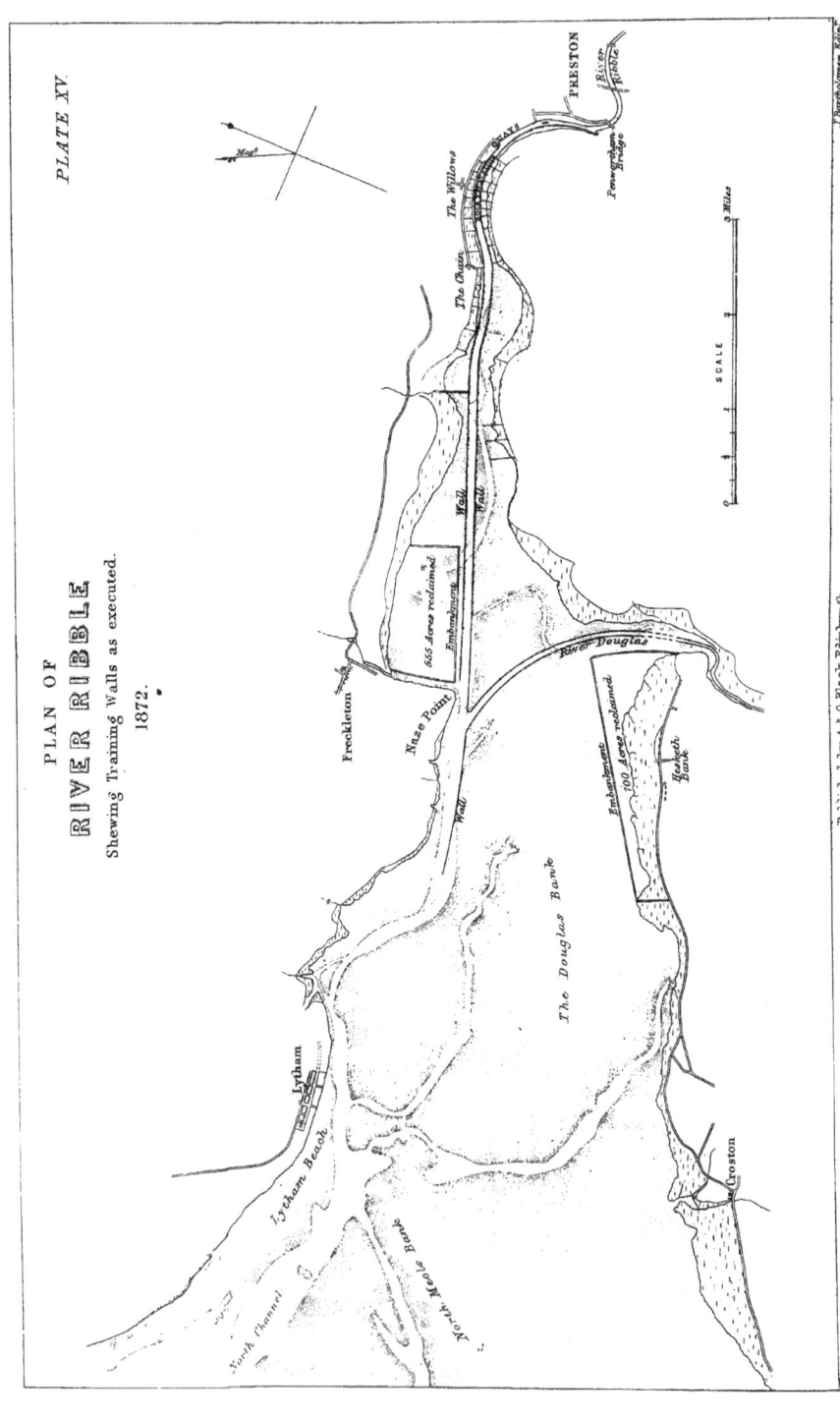

PLATE XV.

PLAN OF
RIVER RIBBLE
Shewing Training Walls as executed.
1872.

Published by A.& C. Black, Edinburgh.

2*d*, That an increase of slope is followed by a retardation of the rate of propagation.

3*d*, That the rate of propagation does not bear any constant relation to the amount of slope, although it is to some extent modified by it.

4*th*, That while the rate of propagation in rivers is in some measure due to the depth of water, it is nevertheless influenced by the slope of the surface, the form of the channel, and the obstructions protruding from the sides or bottom.

5*th*, That, if not in all cases, at least when there are steep slopes and shallow water, as at the Dornoch Firth, the level of the crest of the wave must rise to the level of the surface of the water (or perhaps the bed of the river) above it, before a progressive motion takes place; and

6*th*, That, from the difficulty of dealing with so many variable elements, it is impossible, in many rivers, to determine the ruling circumstances which can be held as regulating the rate of tidal propagation.

RIVER RIBBLE.

The Ribble in Lancashire, as shown in Plate XV., the improvements of which were designed by Messrs. Stevenson, presents an example of a great amount of additional depth having been obtained in a comparatively short space of time. That river, according to Mr. Park, who conducted, as resident engineer, the greater part of the works, has a course of 82 miles, and drains 900 square miles of the counties of York and

Lancaster. The formation of its bed rendered the state of the tidal compartment previous to the improvements very defective. The bottom in the lower part of the river consists of loose sand, while that of the upper reach is alternately compact gravel and sandstone rock. About half a mile below Preston, in particular, it was found that a solid ridge of sandstone, extending to 300 yards in length, stretched quite across the channel. Its surface was from 3 to 5 feet higher than the general bed of the river both above and below it, and so prominent an obstruction did it form that the higher parts of the rock were occasionally left dry during the long droughts of summer. The propagation of the tidal wave and free flow of the currents were checked on approaching it, while the power of the tidal and fresh-water scours was in a great measure neutralized, and rendered almost unavailable in keeping open the upper and lower stretches of the navigation; so that its influence in obstructing the river resembled that of a great artificial weir stretching across the stream. In proof of this it may be stated that the ordinary rise of spring-tides at Lytham, which is 12 miles seaward of Preston, is about 19 feet,[1] and that of neap-tides is 14 feet, while at Preston, prior to the operations, the rise of spring-tides did not exceed 6 feet, and neap-tides of 13 or 14 feet rise at Lytham did not reach Preston at all. The removal of the rock which encumbered the bed was naturally viewed as the most urgent

[1] Captain Sir Edward Belcher, while engaged in making the Admiralty Survey of the Ribble, found that on one occasion the tide at Lytham rose 25 feet 7½ inches.

and important work for effecting an improvement in the tidal phenomena and general depth of water. To this, therefore, the Navigation Company first directed its attention, and, as has been noticed in Chapter VIII., succeeded in removing the rock, and further, in dredging many thousand tons of gravel, and erecting about 18 miles of rubble training walls. I have already given some details as to these works, and I have only to add here that they have effected a striking improvement on the navigation. Mr. Garlick informs me that, at "the Chain," below Preston, the level of the low water is now 6 feet 8 inches lower than it was in 1841, before which period the works had begun to show their effect. So that it is safe to conclude that the total lowering of the low-water line is between 7 and 8 feet, and the tidal range has been increased to the same extent, and the tidal propagation, when I had occasion to ascertain it some years ago, was found to have been accelerated upwards of an hour. The practical result of this improvement is that vessels to which the navigation was previously *at all times closed*, can now come up to the quays at Preston with comparative ease and safety, even in neap-tides.

River Lune.

The works on the Lune in Lancashire were executed by Messrs. Stevenson, under the direction of the Admiralty. They consisted in removing fords by dredging, shutting up subsidiary channels, and erecting river walls. Like all rivers flowing through tracts of sand-banks, the Lune was

ever changing its course, and in order to illustrate this I
have shown on Plate VIII. the channel of the Lune in
August 1847, August 1848, and December 1848, taken
from actual survey. The great object of the improvements was by removing obstructions and making training
walls so to regulate the currents as to insure a fixed
channel and a greater depth. Fig. 48 shows the gradual

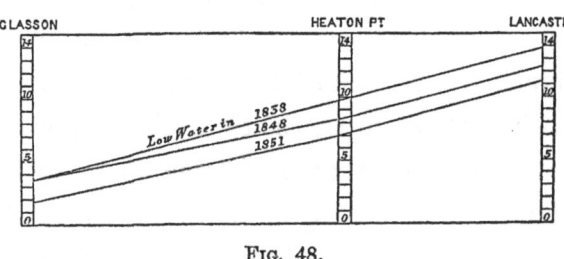

Fɪɢ. 48.

change effected on the low-water line in consequence of
the works. The upper line shows the surface of the
river in 1838, the intermediate line in 1848, and the
lower line in 1851. The general effect has been to
increase the depth of water up to the quays at Lancaster
about 4 feet, and to prolong the duration of the tidal
influence at that place 30 minutes in neap and one hour
and a half in spring tides; so that vessels can approach
and leave Lancaster much earlier than formerly, while the
improved channel is navigated with much greater ease.

RIVER CLYDE.

The Clyde affords a striking proof of the extent to
which river improvements may be carried. So insignifi-
cant was the stream in its natural state, that Smeaton, in

1755, proposed to erect a dam with locks in the lower part of the river, and to convert it into a tidal canal in order to bring craft drawing 4 feet 6 inches up to Glasgow. In 1768, however, Golburne surveyed the river. He found that as far down as Kilpatrick the depth of water was only 2 feet, and recommended the construction of a series of jetties from either side, for the purpose of narrowing and deepening the stream. This may be held to have been the commencement of the improvement of the river Clyde, which now admits vessels drawing 22 feet to steam up to Glasgow. The reader must not however suppose that this result has been attained by means of the jetties which were erected under the advice of Golburne. It was soon discovered that this object could not be gained by such works, and it was not until the ends of the jetties were connected by longitudinal walls, and until dredging machines were extensively employed, that the Clyde improvements began to assume an importance commensurate with the vast commercial interests of the city of Glasgow and surrounding districts. Even so recently as 1836, Mr. James Walker was asked to report to the Trustees on a scheme to construct a canal from Bowling to Glasgow. It is, indeed, since 1836 that the Clyde Navigation may be said to have made its most important progress towards its present state, and this has been achieved in a great measure by widening the river where it had been imprudently contracted, and by dredging on an enormous scale, as I have already stated in Chapter VIII., under the head of Dredging, and nothing now can avail to remove

continually growing obstructions, and to keep the navigation open, but an unremitting application of steam power
applied to the best advantage. It would have been very
interesting, in such a case as the Clyde, which, I may say,
from being a fresh-water stream has been converted into a
great tidal channel, to have possessed accurate records of
the original and present levels of the low-water lines of
the river.

It is not possible to arrive at a correct estimate of the
actual extent to which the low-water level of a river has
been lowered, unless an accurate record has been kept of
the levels with reference to a fixed bench-mark. The
lowering of the low water is a slow process, and the eye
gradually becomes associated from year to year with an
entirely altered state of the river, which, if it had
occurred in the course of a night, or even a week,
would have struck even a casual observer with amaze-
ment. It is, I believe, impossible now to arrive at the
extent to which the bed of the Clyde has been lowered
since the days of Golburne, as the old plans do not specify
a proper datum for reference. But having occasion to inquire into this with reference to a judicial question on
which I was instructed to report to the Court of Session,
I found the means of arriving pretty accurately at the
extent the low-water line had been lowered since 1853
at Erskine Ferry and West Ferry, the former $9\frac{1}{4}$, the latter
$13\frac{1}{2}$ miles from Glasgow; and the following table gives the
results as reported by me to the Court. But even these
figures may be subject to some correction, due to the state
of the river when the observations were made :—

TABULAR VIEW OF THE RELATIVE LEVELS OF LOW WATER AT
VARIOUS DATES FROM 1853 TO 1868.

Levels below
datum.   *West Ferry.*

18·00.   Low water on cross sections, 1853 ; by Thomas Kyle.

18·08.   Low water, spring, 1853, on section ; by Thomas Kyle.

18·17.   Low water, average spring, April to September 1853 ; note
on plan by Thomas Kyle.

18·58.   Low water, spring, 20th March 1840 ; sections by Thomas
Kyle.

19·32.   Low water, spring, 17th September 1868; reported by
David Stevenson.

Erskine Ferry.

17·00.   Low water, spring, 1853, on longitudinal section ; by
Thomas Kyle.

17·15.   Low water, spring, 1853, on cross sections ; by Thomas Kyle.

17·42.   Low water, average spring, April to September 1853 ; note
on plan, by Thomas Kyle.

17·51.   Low water, spring, 20th March 1840, on section ; by
Thomas Kyle.

17·60.   Low water, on contract plan, by John F. Ure, dated 10th
April 1855.

18·18.   Low water in 1866 and 1867.

18·70.   Low water, spring, 19th September 1868 ; reported by
David Stevenson.

Note.—The datum to which the levels refer is the surface of the
cope of South Quay wall of Glasgow harbour, as defined in plan 1853-4,
by Thomas Kyle.

In the upper part of the river the lowering of the bed
has been much greater.   In 1832, the late Mr. Robert
Stevenson erected Hutcheson Bridge, which crossed the
Clyde at the site of the new Albert Bridge.   The masonry
of the piers of Hutcheson Bridge was laid at the level of
7 feet below the bed of the Clyde, on a platform of timber
on piles 18 feet in length.   I found by a section made in

1845, after a lapse of thirteen years, that the level of the river had been lowered, in consequence of the improvements of the Clyde Trustees, no less than 11 feet, and even with that amount of scour the bridge was, and might long have remained, a safe structure. But immediately above its site there is a weir which dams up the Clyde and forms a lake, or almost still pool, for several miles. It was determined to remove this weir, and after its removal the bridge could no longer be pronounced safe; and it has been accordingly replaced by the present structure.

It is right to state, with reference to the removal of the Clyde weir, and as to what I have said on the subject in Chapter VI., that the removal of weirs, viewed as an abstract question, is in general a safe and even proper navigation improvement. But in the case of the Clyde weir, to which I have alluded, two questions were urged by the opponents of the measure. The *first* was, whether the damage to the banks, and the amount of stuff which would be sent down into the lower harbour, would not occasion more expense to dredge it than could be compensated for by any increase of scour due to the removal of the weir? And the *second*, and perhaps more important, question was, whether, as a sanitary and amenity question, it was desirable to sacrifice the lake formed by the weir, and used for bathing and boating by the great population of Glasgow? I thought it was better that the weir should remain; but this is a question more for the community of Glasgow than for engineers to determine.

RIVER TEES.

I shall only further refer to the Tees, as affording an example of a navigation where improvement was long deferred in consequence of advice which was not calculated to attain that object. I am indebted to Mr. John Fowler, of Stockton, the Engineer to the Tees Commissioners, for the information concerning that river. Without going back to its very early history, it is sufficient to say that in 1804 Mr. Chapman found that the available depth up to Stockton was 9 feet at spring tides, and at Cargofleet, five miles below Stockton, there was a shoal with only 2 feet at low water. Following out his report, the Tees Navigation Company was incorporated —the Portrack cut was executed and opened in 1810, and effected some improvement. It does not appear that much more was done till 1827, when the Navigation Company consulted Mr. Robert Stevenson of Edinburgh and Mr. H. Price. Both of these engineers recommended another cut to be formed, as already noticed at page 196, but they differed in opinion as to the general treatment of the river. Mr. Price recommended that it should be contracted by jetties, and Mr. Stevenson that the banks should be faced with continuous walls, stating as his reason for this recommendation, "that to project numerous jetties into the river I regard as inexpedient, being a dangerous encumbrance to navigation, and tending to disturb the currents and destroy the uniformity of the bottom." The plan adopted by the Navigation Company was however that of Mr. Price, and jetties were constructed on the river to a large extent, and Mr. Fowler says "that after a trial of twenty-seven years it was found that they were liable to all the objections that had been urged against them by Mr. Stevenson." Accordingly, under Mr. Fowler's direction, the whole of the jetties have been removed, and the river is now guided between continuous walls in the upper part of its course, and low training walls, similar to those in the Ribble and other rivers, have been constructed in the lower reaches. The result of the operations carried out by Mr. Fowler is that there is now an easily maintained channel having a navigable depth of 18 feet up to Stockton, where, besides the ordinary traffic of the district, there is a large shipbuilding trade, launching steamers of 3000 tons burden.

Many instances might be referred to where a course of treatment opposed to that which I have recommended has not been followed by favourable results; but I deem it sufficient to confine my remarks chiefly to an exposition of the correct principles of river improvement, without discussing at length erroneous practice or its baneful results; the more so as these have been most fully and ably treated by Captain E. K. Calver, R.N., whose investigations into the former and present state of some of our tidal rivers are of great value to the hydraulic engineer.[1]

[1] *The Conservation and Improvement of Tidal Rivers*, by E. K. Calver, R.N. London, Weale, 1853.

CHAPTER X.

It is necessary, however, to state that in certain situations the principles of river improvement which I have advanced will be found to be of very limited application. Such cases, indeed, are rarely to be met with, but still it is necessary to notice them. I allude to rivers where the tidal or intermediate compartments are, from natural causes, of very small extent.

THE ERNE.

To illustrate what is meant, I refer to the Erne in Donegal, which has a tidal capacity of only 2¾ miles, extending from the bar up to the town of Ballyshannon, where the tidal flow is terminated by what is called the "Salmon Leap," a perpendicular bed of rock extending across the river, and rising to a height of 15 feet, over which the river forms a cascade. This waterfall forms the limit of the tidal flow, beyond which it could not, without works of a gigantic character, be extended.

THE NESS.

Another case is the Ness, where, indeed, although there is no waterfall, there exists perhaps a no less seri-

ous obstacle to tidal flow.   The river at present passes by
a bending channel from Inverness to the Beauly Firth at
Kessock Roads, a distance of about 2 miles.   A scheme
was proposed in 1847 for making a straight cut to obviate
the great difficulty which vessels have in making their
way from the Beauly Firth to Inverness, a difficulty which
was mainly attributed to prevailing adverse winds, due
to the configuration of the surrounding hills.   But on
making an investigation, with a view to reporting on the
proposed improvement scheme to the Admiralty, it was
found that the difficulties attending the navigation of the
river are mainly the prevailing outward currents due to the
physical conformation of the bed of the Ness, which may
be shortly described, as it illustrates generally a class of
rivers which are very difficult to improve :—1st, The rise
of ordinary spring-tides at the mouth of the river is 14
feet ; 2d, The distance to which the influence of such
tides extends is only about 2 miles, which comprises the
whole tidal compartment of the river ; 3d, the slope or
inclination of the low-water line of this tidal compartment
is no less than 7 feet per mile, and the tide takes from
2 to 3 hours to make its way up the first mile ; 4th,
The natural result of such a state of matters is, that no
tidal current is generated at the mouth and propagated
up the stream, and consequently the phenomenon of a
current, due to flood-tide, may be said to be almost unknown.

Under these circumstances the main barrier to free
navigation of the river Ness may be traced to the absence
of a tidal current, to aid the entrance of vessels from

Kessock Roads, and assist their progress up the quays. This part of a ship's voyage is at present effected by help of men and horses, which drag the vessel against the nearly constant downward current, which varies in strength with the amount of water discharged by the river Ness, during its frequent heavy floods.

The existence of a moderate amount of fall or slope on the low-water line of a river, is a hopeful feature in its capabilities for improvement; while, on the other hand, such a slope as that on the Ness proves a great barrier to its extended improvement as a tidal river; for it is obvious that to obtain on that river a slope sufficiently gentle for easy navigation, it would be necessary to lower its bed to so great an extent, and to execute works of such magnitude, as to render it inexpedient to entertain such a project.

The two instances I have given will suffice to illustrate those cases, happily not very numerous, which do not come within the range of what I may term improveable rivers, for in either of the cases I have named, works of a magnitude wholly disproportionate to the benefit to be derived would be requisite, in order to remove the obstacle which nature has opposed to the existence of a navigable tidal channel.

CHAPTER XI.

WORKS FOR ACCOMMODATION OF VESSELS.

Docks—Tide-basins—Groynes—River quays; examples of those at Belfast, Londonderry, and the Clyde.

THE works I have already described are for facilitating the ingress and egress of vessels. In addition to this, it is necessary to provide for their accommodation. For this purpose it is desirable, where the currents are strong, to afford them some protection against heavy floods accompanied by ice, which are often very destructive to shipping.

Docks.

On a large scale this protection is afforded by docks entered by tide locks, and constructed in all respects like the wet docks in any of our seaports, and they may therefore be held to be a class of works common to all harbours, and not specially connected with tidal rivers. There are some works, however, that are essentially river works, and these it is necessary shortly to notice.

Tide-basins.

Among them are what are termed tide-basins, which are artificial cuts retiring from the stream, having their sides bounded by quays or wharves, into which vessels may be withdrawn and sheltered from the current, but where they are still liable to take the ground at low water. The Kingston dock at Glasgow is a large tidal basin of this kind, being in fact a dock without locks or gates. But it is sometimes desirable on a smaller scale to protect the berthage of quays along rivers from currents or accumulations of gravel. This was done at Inverness by a weir or groyne of timber work, which is shown in elevation and cross section, fig. 49. It extends in front of

Groynes.

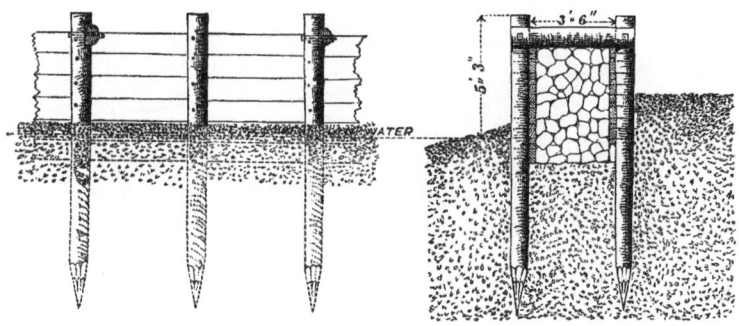

FIG. 49.

the quays, and prevents the current of the river in floods from shoaling the berthage by heaping up gravel. I have seen a protection on a very large scale at Albany, on the Hudson, where the vessels navigating that river, and trading with the Erie Canal, are accommodated in a large basin of thirty-two acres. This basin is separated from the current of the Hudson, and the ice it sometimes brings down, by a longitudinal mole or pier of about three-quarters of a mile in length, left partially open for scour at the upper end, and connected to the shore by drawbridges.

In many situations where currents are not very strong, and the river is sufficiently wide to admit of vessels being moored in it, as at the Clyde, or the Foyle at Londonderry, the berthage for vessels is very conveniently

River quays.

afforded by forming lines of quays along the shore. Such quays, indeed, constitute an important part of all harbours which are formed in tidal rivers; and in illustration of some of the various methods of construction adopted in such cases, I may give the following cross sections. Fig. 50 shows the timber wharfage constructed by Mr. Smith

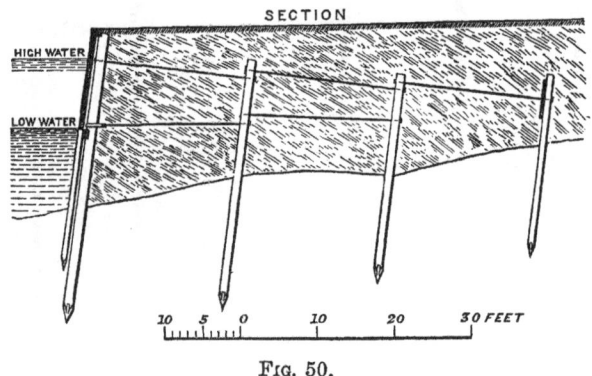

FIG. 50.

at Belfast, which is composed of a facing of timber-work secured by iron ties fixed to piles, the space behind the framework being filled up and the roadway formed at the top. Fig. 51 is a plan showing the positions of the piles

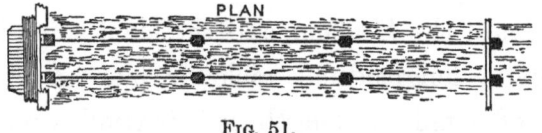

FIG. 51.

and ties. Sometimes a similar face-work is employed, backed by a wall of concrete, and iron plates have also been used for the facing instead of planking. Figs. 52 and 53 are a section and elevation of the quays at Londonderry, designed and executed by Messrs. D. and T.

Stevenson.   At this place the ground is very soft, and in order, as much as possible, to reduce the weight, the front

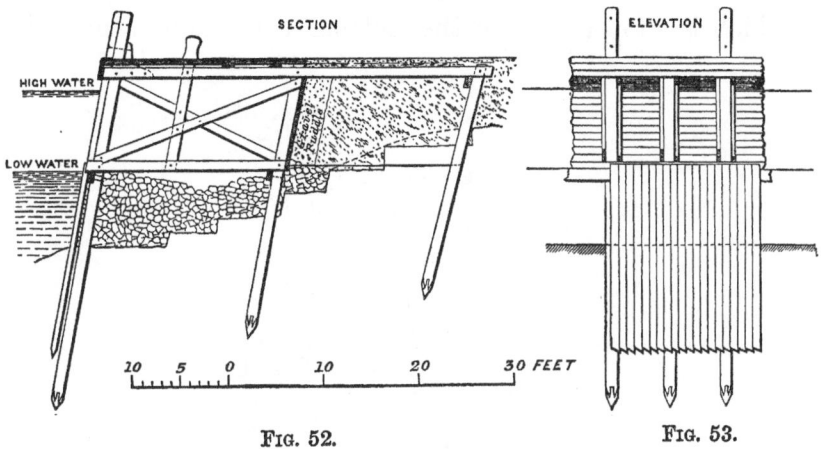

FIG. 52.   FIG. 53.

compartment of the wharf next the river is left open.

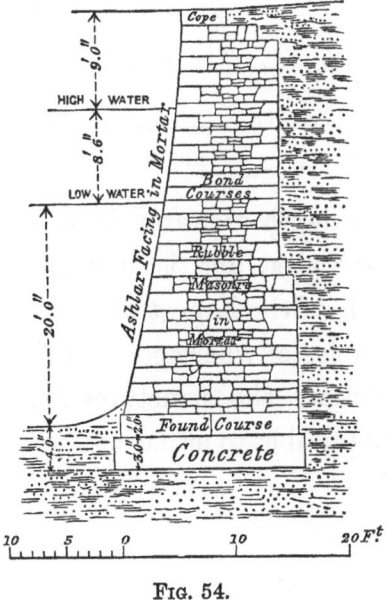

FIG. 54.

Figs. 54 and 55, again, are sections of the stone wharves, constructed from a design by Mr. Walker, at Glasgow, under the superintendence of Mr. Ure.　Fig. 54 is the section adapted to a clay bottom; and fig. 55 is that which is adopted when the bottom consists of sand.　In

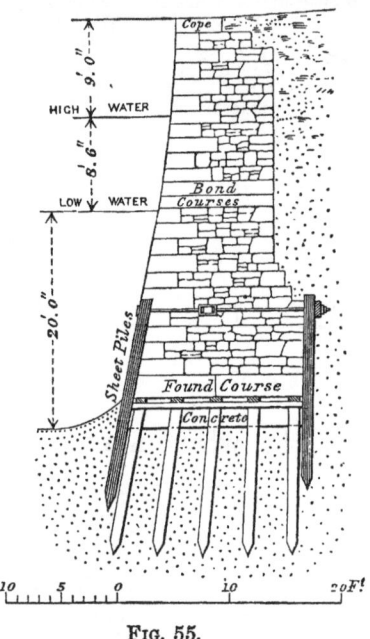

FIG. 55.

both cases the depth of water in front of the quays is 20 feet at low water, and is intended to accommodate merchant vessels of the largest class.

These examples of wharfage on tidal rivers will serve to show the student the structures designed by different engineers as applicable to situations where the foundation is hard or soft.

CHAPTER XII.

" SEA-PROPER " COMPARTMENT OF RIVERS.

Bars ; theories to account for the formation of—Origin of bars, as illustrated by the Dornoch Firth ; Castelli's theory of the formation of bars ; conditions under which bars are formed—Barless rivers—Bar at Cochin—Depth over bars due to scour—Comparison of river and tidal water in estuaries—Backwater ; its importance for scouring ; different aspects under which backwater may be viewed, as illustrated by Hartlepool slake, Montrose basin, and Wallasey Pool—Level at which backwater is abstracted—General propositions regarding backwater—Lower parts of estuaries, such as the Mersey, etc., cannot be improved unless at great cost—Bars of such rivers as the Tyne and Wear may be improved by protecting piers—Bar of the Mississippi—Bar of the Danube ; its cause, and works for its improvement—Hard bars—Groynes.

MANY of the works described in Chapter VIII., such as training walls and dredging, are not more applicable to the "tidal" than to the "sea-proper" compartment, the *distinguishing* features of which, are the phenomena attending the flow of rivers or bodies of tidal water into the sea.

BARS.

In some instances, such, for example, as the Forth, the junction of the estuary with the sea occurs without occasioning any very perceptible or marked disturbance of the currents or change on the bed of the channel, so that a ship may, at any time of tide, run without check or hindrance from the Isle of May to St. Margaret's Hope. But in this respect the Forth is exceptional.   The entrances to almost all British, as well as Continental, rivers are interrupted by what is termed a "bar." Its origin is, indeed, not always to be traced to the same cause, but the Mersey and the Tay, as well as the gigantic Nile and Mississippi, have the same troublesome feature, which is not only very hurtful to navigation, but is perhaps the most difficult subject with which the marine engineer has to grapple.

A "bar," in nautical language, is the name applied to the shallowest part of the navigable channel through the sand-banks which generally collect at the mouths of estuaries. It may perhaps best be explained by an

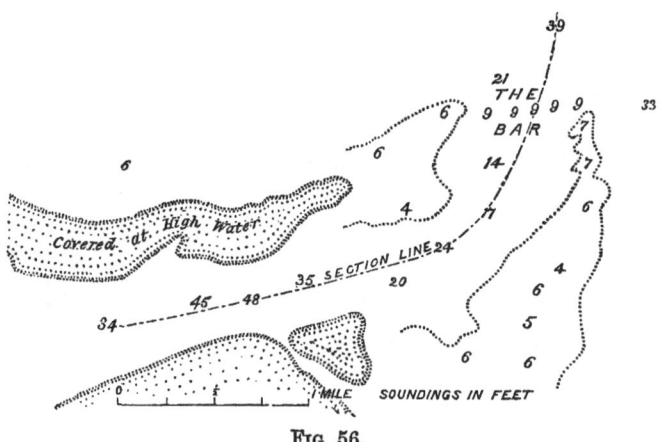

FIG. 56.

illustration. Fig. 56 is a diagram, on which the dotted line shows the fairway or deepest channel through the sand-banks. It will be seen that the place marked "the bar" lies at a considerable distance from the shore, and has extensive sand-banks, *drying at low*, but covered at high, water, as well as submerged sandbanks which *never dry*, on either side of it. Fig. 57 is a longitudinal section

made on the dotted line, and represents the depth of
water on an enlarged scale.   From this it will be under-

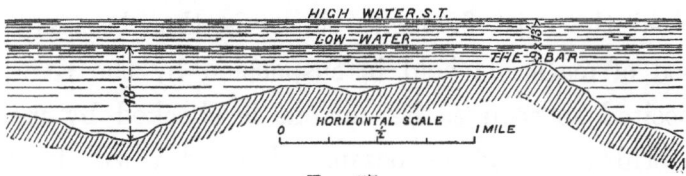

FIG. 57.

stood that the bar is the shallowest part of the channel,
there being deep water both landward and seaward of
it.   What is termed "the bar," therefore, is not the sand-
bank that dries at low water, but those *constantly submerged* banks which have a channel, subject to variations
in position and depth, passing through them.

The bar then regulates the navigable depth, and no
passage over it can be obtained until the tide has risen
sufficiently high to enable vessels to cross it ; and it is
more or less marked and decided according to certain
conditions to be afterwards explained.   We accordingly
find great variety in the depth of water.   For example,

The bar of the Mersey has a depth of from 9 to 10 feet at low water.

„	Tyne	„	6 to 7	„
„	Wear	„	3 to 4	„
„	Ribble	„	7 to 8	„
„	Tay	„	16 to 18	„
„	Dee	„	10 to 12	„

And while these limited depths exist on the bar,
there is in most cases ample depth within or landward
for vessels of the largest class to lie afloat at all times
of tide.   At the Dee, for example, the celebrated anchorage of "Mostyn Deep," affords depth and area for almost
any fleet of ships.

Many theories have been propounded to account for the phenomenon of the " bar." What may perhaps be termed the most favourite theory is, that bars are composed of materials held in suspension by the river, and deposited so soon as its current is checked by meeting the still water of the ocean. This idea will be found stated by various authorities, as expressed in the following quotations :—

" When the flood matters meet the incoming tide, there must necessarily be a deposit."

" The position of a bar is at the point where the opposing forces meet or balance. The material held in suspension by water, travelling at a certain velocity, falls to the bottom and forms a deposit where that velocity is checked."

" The incoming tide, when it meets the water discharged by the river, checks the velocity of this water, and so causes a deposit, which forms the bar."

Many other similar quotations could be given. But this theory, at all events as regards " sea bars," of which we are now treating, is disproved by such a case as the Dornoch Firth. The bar at that place occurs at a point 14 miles seaward of the point at which the river enters the sea, as will be seen in Plate IV. The idea that sand-banks of such magnitude as those at the Dornoch Firth could be formed by the detritus brought down by the small rivers Oykell and Cassley, which flow into the upper end of the firth, is wholly untenable, and is indeed contradicted by the fact that the bar and adjoining banks are composed of pure sand, and not of alluvial matter deposited by the river, as will afterwards be more fully alluded to.

Another theory attributes bars to the want of sufficient scouring power; but this as an abstract statement is unwarranted when we find bars existing at the mouths of such rivers as the Mississippi.

Another theory attributes the *absence* of a bar to "the presence of a nearly equal duration of the period of ebb and flow in the lower reach of the river, accompanied by an extremely gentle inclination of its surface at low water."[1] To refer again to the Dornoch Firth: we have an equal duration of the ebb and flow throughout the firth, and the level of low water practically the same; and yet there exists as perfect a specimen of a bar at the mouth of the firth as can possibly be imagined. We cannot, therefore, in endeavouring to account for the existence of bars, or the exemption from them, accept any of these explanations.

Since 1842, when I had occasion to bestow much attention to the subject, I have never had any difficulty in tracing the accumulations which give rise to all such bars as those of the Mersey, the Ribble, the Tyne, the Dornoch, or the Tay, subject to the conditions hereafter stated, *solely* to the action of the sea. The waves, as is well known, throw up a girdle of light or heavy material, varying with the exposure from sand to boulders, round every bay and headland of our coast; and the entrances to rivers form no exception. The effect of this constant action of the sea is to form a continuous line of beach

<hr>

[1] *Treatise on the Improvement of the Navigation of Rivers*, by W. A. Brooks.

across the mouths of all our tidal rivers and inlets; and such a beach would very soon be made, did not the flow of the tidal currents maintain an open channel through it. In this way the waves of the sea on the one hand, and the currents on the other, produce the well-known feature of a tide-covered beach extending from the shores of our inlets, with submerged sand-banks, having a channel through them termed the "bar."

This explanation is given in a report made in 1842, in which it was necessary to investigate the cause of the "bar" at the Dornoch Firth, from which the following is an extract:—

Bar at Dornoch Firth.

"This bar is an accumulation of hardish sand-banks, through which there is a navigable fairway of not less than 9 feet at low, and 22½ feet at high water, of ordinary spring-tides. It appears to be retained in its present state by a combination of agents. The heavy swells from the German Ocean, with which the coast is visited, have a tendency to heap up the sand from the adjoining shores; but this tendency is, to a certain extent, counteracted by the tidal and fresh-water currents of the firth, and the result of their joint action is the bar—a bank, or series of banks, of considerable extent, permanently under water, through which there is a deeper passage or fairway, whose depth of water is believed to remain pretty much the same, although its direction occasionally changes."

But I cannot altogether claim to be the author of this explanation of the origin of bars, as I afterwards discovered that a suggestion in some respects the same had

been given about two centuries ago by the Abbot Castelli, Castelli's theory
who wrote as follows :[1]—"As to the other point of the of bars. of the formation
great stoppage of ports, I hold that all proceedeth
from the violence of the sea, which being sometimes
disturbed by winds, especially at the time of the waters
flowing, doth continually raise from its bottom immense
heaps of sand, carrying them by the tide and force of
the waves into the lake; it not having on its part
any strength of current that may rise and carry them
away, they sink to the bottom, and so choke up the ports.
And that this effect happeneth in this manner, we have
most frequent experience thereof along the sea-coasts;
and I have observed in Tuscany, on the Roman shores,
and in the kingdom of Naples, that when a river falleth
into the sea there is always seen in the sea itself, at the
place of the river's outlet, the resemblance, as it were, of
a half-moon, or a great shelf of settled sand under water,
much higher than the rest of the shore, and it is called in
Tuscany *il cavallo*, and here in Venice, *lo seanto;* the
which cometh to be cut by the current of the river, one
while on the right side, another while on the left, and
sometimes in the midst, according as the wind fits.   And
a like effect I have observed in certain little rillets of
water along the Lake of Bolsena, with no other difference
save that of small and great."

Had the Abbot ended his statement here, it would have
been identical with that I have suggested, but he goes

---

[1] *The Mensuration of Running Waters.*   By Don Benedetto Castelli, Abbot of
St. Benedetto Aloysio, and Professor of the Mathematics to Pope Urban VIII. in
Rome ; translated by Thomas Salusbury, Esq , London, 1661.

on to say, "Now whoso well considereth this effect plainly seeth that it proceeds from no other than from the contrariety of the stream of the river to the impetus of the sea-waves; seeing that great abundance of sand, which the sea continually throws upon the shore, cometh to be *driven into the sea by the stream of the river*, and in that place where these *two impediments meet with equal force, the sand settleth* under water, and thereupon is made that same shelf or *cavallo;* the which, if the river carry water, and that any considerable store of it shall be thereby cut and broken, one while in one place, and the other while in another, as hath been said, according as the wind blows; and through that channel it is that vessels fall down into the sea, and again make to the river, as into a port."

The words which I have italicised speak of sand "driven into the sea by the stream of the river," and of the place where the sea and the river meet with "equal force," causing the sand "*to settle*," and are at variance with the suggestion I have proposed. In the cases to which my explanation refers there is no settlement on the bar of sand, or other material carried down by the river. Neither is it necessary to the formation of a bar that there should be "a place where the river and sea meet with equal force," so as to cause sand held in suspension to settle. For, according to my explanation, a sea-bar would be formed although the outgoing current held not one particle of matter in suspension, its only effect being to scour away what the waves have thrown up.

After having given much attention to the subject of *Conditions under which sand-bars are formed.* the sand-bars which encumber most of the tidal harbours of the shores of Britain, I proposed, in the *Encyclopædia Britannica*, the following as the conditions under which all such accumulations are formed :—

1st, *The presence of sand or shingle, or other easily moved material;*

2d, *Water of a depth so limited that the waves during storms may act on the bottom;* and

3d, *Such an exposure as shall allow of waves being generated of sufficient size to operate on the submerged materials.*

In confirmation of this opinion, I may once more refer to the Dornoch Firth.   The Oykell, as has been shown in Chapter III., joins it at a point about a mile below Bonar Bridge, but we find no indication of what may be termed a bar throughout the whole of the sheltered part of the firth, which extends for 12 miles seaward of that point, until we reach the outer portion, where, open to the whole fetch of the Moray Firth, there are generated waves of sufficient size to act on the materials of which the bottom is composed, and we find an extensive sand-bank, forming, as it were, a continuation of the shore on either side, and stretching quite across the mouth of the firth, with the bar in the centre of it.

The same reasoning may explain why, in such a case *Barless rivers.* as the Firth of Forth, for example, no bar exists.   The Firth of Forth is an inlet or arm of the sea of great width and depth ; the seas entering it do not act on the bottom so as to disturb and heap up the material of which it is composed, in the same manner as in a shallow sea. This great natural depth continues as the Forth gradually contracts; and before the necessary conditions for the formation of a bar occur, namely, shallow water and presence of sand or other easily moved material, the sea is so land-locked that waves of sufficient size to produce the necessary effect cannot be generated. There is, in fact, in the Forth that *gradual diminution of depth and increase of shelter which combine to produce the phenomenon of a river without a bar.*

Bar at Cochin.   It is very interesting to know that Mr. George Robertson, in his recent survey of Indian harbours,[1] found that at Cochin the removal of certain projecting spits of sand which protected the bar had sensibly reduced the depth of water, as ascertained by actual survey, thus affecting, by the operation of changes wrought by nature, a striking proof of the soundness of the condition which I have specified as being necessary to the formation of a bar. Reporting on Cochin, he says,—"Were the current kept together till it got into such a depth of water that the action of the waves was not sufficiently powerful to stir up the bottom, there would be no decided bar. The same result would happen were the current to discharge under shelter from the waves; as, for instance, in a land-locked estuary. In the survey of 1835, when the current was kept longer together by more projecting *fauces terræ*, and by hard sand-banks, which prevented the stream from spreading, there were

---

[1] Reports to the Government of India on Indian Harbours, by George Robertson, Civil Engineer, 1871.

16 to 17 feet on the bar. Since then, the *fauces terræ* have been gradually eaten away by encroachments of the sea ; and the survey of 1852 shows that a bar was beginning to form with only 13 feet on it. In 1858 the bar had completely formed, but was very narrow. These surveys illustrate the true theory of the formation of bars at river mouths more beautifully than any set of surveys with which I am acquainted ; for it seldom happens that the *fauces terræ* of a river are so much eaten away, and the results of their diminution so plain."

A strong argument in favour of the explanation I have proposed is to be found in the fact that these bars are invariably in their shallowest state after heavy seas. *Bars shallowest after heaviest seas.* This view is also borne out by the material of which they are composed. I have examined with care the deposit at the mouths of many such estuaries, and I have invariably found that the bar and outer banks consisted of coarse-grained sand, without a particle of alluvial matter, which, as I shall have occasion afterwards to notice, is confined to the inner bed and banks of the river.

Indeed, when it is kept in view that the river water floats on the heavier salt water of the sea, and that the current on the bar is invariably stronger than on the shallow sand-banks on either side, it is impossible that the light matters held in suspension by the river can "settle down" or "deposit" on the bar. On the contrary, they are swept out by the rapid ebb current, and, as has been already mentioned; can often be traced for a considerable distance out to sea.

In 1852 Mr. Meik made a series of very interesting *Experiments at the Wear.*

experiments at the mouth of the Wear, to ascertain whence the material dredged from and deposited on the bar had been brought. The dredgings consisted of sharp, gritty sand, brickbats, chalk, flints, pebbles, and marly rock materials, precisely of the same character as had been deposited on the beach adjoining the entrance to the harbour. In order to test this, a number of small billets of wood loaded with lead were deposited at various points, and these were gradually moved by the action of the waves towards the harbour's mouth, and Mr. Meik states that there is only one inference to be drawn from the experiment, and that is, " that by the agency of the flood-tide, ballast and other material has been swept from the east foreshore of the south dock to the harbour mouth, and there settled in the deep water channel, to the prejudice of the bar."

In open bays, in extreme exposures, such, for example, as Wick in Caithness, no indication is to be found of a sand-bank across its mouth, the violence of the waves prevents its formation, and the whole bottom of the bay becomes a submerged beach.

Depth over bars due to scour.

From what has been said, the reader will at once see that the depth of water on such bars as are caused by the waves of the sea, is due to the scour produced by the tidal currents, which cross them four times in every twenty-four hours. These two agents, the *waves* and the *tidal scour*, are constantly opposed the one to the other, and the general principles which should guide the engineer in all designs for increasing, or even maintaining, the depth upon sea-bars, is the preservation of a suffi-

cient amount of tidal water to counteract the tendency of
the sea to heap up detritus at the mouths of our harbours.

That the beds of the upper parts of rivers are scoured,
and their depth maintained by the flow of the *fresh-
water* stream, is not to be questioned; and it is also
beyond doubt, that in many situations the upper portions
of the tidal compartments of rivers are kept open in a
great measure by the *fresh-water* stream, as shown at
page 192; but it seems to me to be no less certain that
the opinions which would assign the depth of water in
the lower parts of tidal rivers, and also through estuaries,
to any other cause than the action of *tidal* water as the
chief agent, are erroneous. In proof of this, I think, I
have only to refer to some of the investigations which have
from time to time been made to ascertain the amount of
the river or fresh water, as compared to the volume of the
tidal water of some of our firths and estuaries.

By means of a series of careful observations and
measurements made at the Cromarty Firth in 1837, to
which reference has already been made, Mr. Alan Stevenson
found that the river Conon, when highly flooded (a state
of matters which of course occurs only occasionally), discharges during twelve hours a quantity which is only equal
to $\frac{1}{57}$th part of the water which passes out of the firth at
every ordinary spring-tide, and $\frac{1}{29}$th of that which passes
out at neap-tides. In its summer-water state, the produce
of the river is reduced to $\frac{1}{1541}$ of the discharge of the
firth in spring, and $\frac{1}{791}$ of the discharge in neap-tides; a
quantity too small to affect appreciably either the velocity
of the currents of the firth or their scouring power. It

Comparison of river and tidal water in estuaries.

has often been argued, that in situations where the velocity of the ebb exceeds that of the flood-tide, the excess
is due to the increased quantity of water passing out with
the ebb, the volume of the ebbing waters being assumed
to be augmented by the amount discharged by the river.
But this is wholly disproved in the case of the Cromarty
Firth; for while the increased quantity due to the river
is seen to be only from $\frac{1}{791}$ to $\frac{1}{1541}$, the average velocity
of the flood-tide at that place was found to be 2·9 miles
per hour, while that of the ebb was 3·6; an increase
which is in all probability due to the tide beyond the
Suters falling more rapidly than it rises, and thus producing a greater head and more rapid current on the ebb,
or to some action of the under-currents which have been
stated to exist there, but is assuredly not due to any
augmentation of water from the discharge of the Conon.
The Tay presents another example of the disproportion
between the tidal and river waters. That river, as gauged
by Mr. Leslie when in flood, was found, including the
Earn, to discharge 969,340 cubic feet per minute. Mr.
Walker, in his Report to the Trustees of Dundee Harbour, assumes the discharge in round numbers at one
million cubic feet per minute, or 240,000,000 during four
hours, and arrives at the following conclusion :—" To
compare the above with the effect of the tidal water at
Dundee, I assume 15,000 acres as the average area (above
Dundee) of the reservoir or estuary during the first four
hours of the ebbing tide, and the vertical fall of tide
during these four hours to be 11 feet. This will give
7,187,400,000 cubic feet, or thirty times the 240 millions

of river water. To compare the effect upon the *bar*, the area of the river between Dundee and the bar must be added ; and the tidal water upon the bar will be upwards of *forty times* the river water," and this, it should have been added, only at the exceptional times when the rivers are in high flood. One other example may be given to show the disproportion between the areas of the inner and outer channels. At the Dornoch Firth the high water area of the channel, at Bonar Bridge, is 459 square yards ; at Meikleferry it is 9047, and opposite Whitness Point it is 25,183 square yards, being fifty-five times greater than at Bonar Bridge.

BACKWATER.

But to the effect of the sea-waves to collect, and the tidal scour to remove, sand-banks, may be traced the origin of a very important question, which has occasioned much discussion and difference of opinion among engineers, and may be stated as follows :— Within the bars of all rivers or firths there is a certain expanse or area over which there flows at every tide an amount of tide-water measured by the extent of the area, and the depth to which it is overflowed. The water so impounded at high-tide is what is called "back-water," a term due, no doubt, to its passing back to the sea over the bar, and the question to which I have alluded as having so much engaged the attention of engineers, is, how far this area occupied by backwater may be en-croached on by solid works displacing the water, without injuriously affecting its scouring power on the bar and lower reaches of the river. At first sight it might seem safe to pronounce that no occupation of tide-covered space can be made without prejudicially affecting the scour, but after a little more inquiry we shall see that this is not strictly the case.

Different aspects under which back-water may be viewed. Now this question of backwater presents itself to the engineer in very different aspects, as modified by the varying physical features of different localities, and perhaps, in treating of it, I shall most satisfactorily illustrate its bearing on navigation by referring briefly to some controverted cases in actual practice, which were argued wholly on the question as to whether, in consequence of certain works, and under certain physical conditions, backwater might be excluded without prejudicially lessening the scouring power. The cases selected illustrate, to a certain extent at least, the different aspects under which the engineer may be called on to view the question, and I do not doubt that other illustrations will occur to other engineers, founded on their own experience.

At Hartlepool Slake. The first example to which I shall refer is Hartlepool. Immediately above the harbour there is a tide-covered area of 173 acres, called the " Slake," communicating with the harbour by a narrow entrance. The whole of the water, which at every flood-tide pours into, and at every ebb flows out of, this vast natural basin, passes over and scours the sea-entrance into the harbour of Old Hartlepool. The harbour authorities placed gates across the entrance to the Slake, and, in order to render the scour more effective, impounded the water at high-tide, and at low water allowed it to escape through large sluices formed in the gates so as to act upon the harbour between half-ebb and low water, at which period the scour is found to be most efficacious. A proposal was made by a company seeking powers under an Act of Parliament to enclose a portion of the area of the Slake, and I was appointed to report to the Admiralty on the propriety of sanctioning the encroachment. After full inquiry I had no difficulty in advising that the proposed encroachment would decrease the scouring power, because it was proved in evidence that when the gates were left open the high-water mark in the Slake, and that in the outer harbour beyond the sluices, attained exactly the same level, showing that the basin was not too large to contain all the water that could be supplied by the flowing tide, and therefore, that it was not safe for the Harbour Trustees to part with any portion of the tide-covered area. The effect of closing the tide-gates and permitting the Slake to be filled by the sluices was also stated in evidence, and the result is interesting and important. It appears that when the tidal flow into the Slake is checked by shutting the gates, and the only supply is made to pass through the sluices, their water-way is not sufficient to fill the basin, and the high-water level does not, in that case, reach within four inches of the natural tidal range outside, so that a quantity of water, amounting to upwards of 90,000 cubic yards, is excluded when the Slake is filled through the sluices.

The other case to which I shall refer is the tidal basin at Montrose, which will be found to present a totally different tidal action. The basin at Montrose has an area

At Montrose Basin.

of 1200 acres, and, like that at Hartlepool, is a natural reservoir which scours the sea-channel of the harbour. From careful observations made by Mr. George Buchanan, it was found that on an average of tides the high water in the basin is upwards of 9 inches *below* the level of the high water outside, indicating that the flood-tide does not flow sufficiently long to fill the basin ; and from this fact it was assumed that a proposed embankment, which had the effect of reducing the area of the basin, might be sanctioned without injury to the scour, as the only result would be to cause the water displaced by the embankment to spread itself over the surface of the basin, slightly raising its level, and thus compensating for the portion abstracted. With reference to a portion of the water this is no doubt true, but it would not be safe to carry this assumption beyond a certain limit, for, as suggested by Mr. Buchanan, though the level of the water in the basin be raised by water which formerly occupied the space enclosed by the embankment, it must not be overlooked that the velocity of the current flowing into the basin will be reduced in proportion to the reduction of head between the surfaces of the water *within* and *without* the basin, so that the abstraction of backwater in such a case must depend on whether there be time for the basin to fill with the reduced head.

At Wallasey Pool.

Hartlepool and Montrose are basins into which, as I have explained, the sea ebbs and flows ; but there are other cases connected with tidal rivers in which the " backwater " question forms an important element : for example, Birkenhead Dock on the Mersey. The scheme for that work, designed by the late Mr. J. M. Rendel, contemplated a displacement of 3,750,000 cubic yards of tide-water from Wallasey Pool opposite Liverpool, covering an area of nearly 300 acres. This was opposed by the Liverpool Dock Commission on the alleged injury that the abstraction of so much water would produce on the bar. The promoters of the Wallasey Pool scheme contended that no abstraction of water would take place in consequence of the wall they proposed to erect across the mouth of Wallasey Pool, and averred that the water which formerly flowed into the pool would, after the erection of the wall, flow into the upper part of the river, and be as effective as ever in scouring the bar. This averment was based on the result of tidal observations which showed that the wide expanse caused by Wallasey Pool produced a decrease in the velocity of the tidal currents, and a depression in the level of the water opposite the pool. The observations also showed that, notwithstanding the disturbance of tidal flow caused by Wallasey Pool, the water moved up the estuary with a momentum which raised the level of high water at all the stations in the upper part of the river. Thus taking Princes Basin, nearly opposite Wallasey, as zero, the *means* of the heights of spring and neap tides at different points were as follows :—

	Height at Springs. Below zero.		Height at Neaps. Below zero.	
	Ft.	In.	Ft.	In.
Princes Basin,	0	0	0	0
Ellesmere Port,	1	1	0	8
Runcorn,	1	1	0	11
Fidlers-ferry,	1	8	0	10
Warrington,	2	3	1	6

It further appeared from calculation that the raising of the level of that part of the estuary which lies above Wallasey Pool to the extent of 1·14 inch would give an amount of water equal to the whole quantity displaced by the closing of the pool. After considering all the data adduced, I came to the conclusion arrived at by Mr. Rendel and the other engineers who supported the Bill, that the wall, if built in the line proposed, would regulate the current, restore the lost momentum opposite to the pool, and cause more water to pass into the upper reaches of the river, and that on the whole the scour on the bar would not be appreciably affected. After a contention of twenty-four days before Committees of both Houses of Parliament in 1844, the Bill was passed, and the wall has since been made, and Mr. Lyster, the present engineer to the Liverpool Dock Commission says—"The abstraction of water by the construction of the Birkenhead Docks had had no effect upon the bar of the Mersey, although at one time it was thought that the loss of so considerable an amount of water as that from Wallasey Creek would affect the condition of the entrance channels to the river, but the depth over the bar remained the same as it was fourteen or fifteen years ago."[1]

Level at which backwater is abstracted.

Another important question as affects scour is the level at which "backwater" is abstracted. The abstraction of water from a marsh on a high level covered only at high spring-tides is very different from abstracting an equal amount of water from a space which is filled by every tide.

[1] *Minutes of Proceedings of the Institution of Civil Engineers*, vol. xxvi. p. 425.

It will readily be seen that the efficiency as a scour of
a cubic yard of water filled and emptied by *every tide,* as
compared with that of a cubic yard filled only *five times
during every set of spring-tides,* is in the ratio of 730 to
144, not to mention the more effective scouring power
of water discharged after half-ebb, as compared to a
similar quantity discharged, for example, during the first
hour after high water.

The value of the water as a scour is therefore influenced both by its *volume* and by its *level,* and may be
expressed as follows :—

$$S \propto V\,T,$$

where V = the volume or cubic feet of water space above the low-
water level of the estuary.

T = the number of times it is filled by the tide throughout
the year.

S = the effective scouring power.

The only other consideration that should be kept in
view is that of two spaces, V, V, of equal capacity, and
filled *every tide,* that which is lowest in position will be
most effective in operating on the low-water channel.
These values must of course be held applicable only to
different conditions of the *same river* where the hardness
of the bottom to be scoured and other circumstances
remain unaltered.

The different examples I have given will serve to
illustrate the general principles on which almost all
"backwater" questions are treated, and from what has
been said it will be apparent that every new case
that occurs must be regarded with a special view to
its own distinctive features, as suggested by physical

elements peculiar to each locality, such as the configuration of the banks and bed of the estuary, the simultaneous levels of the surface of the water at different periods of the tide throughout the estuary, the velocities of the surface and under currents at different periods of tide and the times of ebbing and flowing, together with many other more minute data *peculiar to each case,* which it is not possible to specify in a general summary.

General propositions regarding back-water.

Perhaps, however, the following general propositions, if not in all cases applicable, may nevertheless be held to represent pretty accurately our general knowledge as regards " backwater :"—

1. *The depth on Bars is due to backwater.*

2. *Where the high-water level of the surface of the river, estuary, or basin is the same as, or higher than, the level seaward of the point of abstraction, a diminution of tide-covered area will reduce the effective backwater.*

3. *Where the high-water level of the surface in the river, estuary, or basin is lower than the level seaward of the point of abstraction, a diminution of tide-covered area may, in some cases, be made without reducing the effective backwater.*

4. *The lower the level of backwater the greater will be its effect in scouring the low-water channel, and, therefore, the nearer the site of abstraction is to high-water mark the less injurious will be the effect.*

5. *By enlarging the tidal capacity of a river at a low level, where the acquired volume is filled every tide, compensation may be given for a much larger amount of water excluded at a higher level.*

6. *In consequence of the disturbing effects of the waves of the sea, the large discharge of rivers during high floods, and the varying nature of the beds of estuaries and bars, it is not possible to conclude that with a given quantity of backwater, as deduced from the measurement of the tidal capacity of an estuary, a constant navigable depth can be maintained over the bar.*

In many of the navigable rivers in this country, such as the Mersey, the Ribble, or the Tay, the lower part of the estuary presents the feature of large tracts of sand-banks, some covered to a small depth and others drying at low water, and the bar, which we have been considering, is situated far to seaward—at the very outskirts, if I may so express it, of these accumulations. It is not a little remarkable that in such circumstances the position of the bar and the depth of water upon it, though varying from time to time, as affected alternately by summer calms and winter storms, should on the whole maintain for years, if not the same position, at least pretty much the same *average* depth of water, and, indeed, that the variation either in position or depth should not be such as materially to incommode navigation, much less to close the access to the harbour. We find, for example, that the entrances to Liverpool, Dundee, and many smaller harbours, although across bars and through extensive sand-banks, have always had a pretty uniform depth maintained by the scour of the backwater which keeps the channels open.

Little has been attempted to improve the entrance to such estuaries by artificial works, partly no doubt from

Lower parts of estuaries such as the Mersey, etc., cannot be improved unless at great cost.

the great expense that would attend any such operation, and partly on account of the difficulty in predicting what effect such works might have on the tidal currents, in situations so exposed to the sea, and how far any interference with their flow might prove beneficial or detrimental.   The tendency has rather been in such large estuaries to trust to the natural scour of the tidal waters, and by careful *lighting* and *buoying* to indicate to vessels the navigable track by following which the mariner will find sufficient water at the proper time of tide to float his vessel over the bar and carry her to her destination.

Changes on banks at the Tay.
The changes which take place in the sand-banks and bars of such estuaries as those to which I have been referring are capricious, and in many cases unaccountable. For example, it is shown by existing surveys that the position of the bar at the mouth of the Tay was the same in 1689, 1816, 1833, and 1846; but a survey made in 1858 showed that it had shifted a little to the north, and what was from the earliest times known as the navigable channel no longer had the deepest water.   Throughout all the periods mentioned, however, the bar had, and still has, a navigable depth for the largest vessels.

At the Mersey.
At the Mersey the changes in the position of the bar have been more frequent, but from observations made by Captain Hills, the Marine Surveyor to the port, it does not appear that they have been accompanied by any permanent diminution of average navigable depth, nor, indeed, by any change in the general level or area of sands dry at low water, or in the tidal phenomena of the estuary.   What was called "The New Channel" of the Mersey, discovered

in 1833, continued navigable for six years, and within eight years of its discovery became obliterated. The second, or "Victoria Channel," was buoyed in 1839, and after gradual deterioration was disused in 1857, and superseded by the "Queen's Channel." Referring to observations on the banks and tides, Captain Hills states the area of sand-banks dry at low water as follows:—

In 1735-6,	.	.	.	= 27·97 square miles.
„ 1833-5,	.	.	.	= 27·82 „ „
„ 1857,	.	.	.	= 27·06 „ „

The mean height of high water above the Old Dock Sill he found to be as follows :—

1st. *Mean Height of High Water throughout the Year.*

				No. of tides observed.
1768,	.	.	15·510 feet above O. D. S.	703
1769,	.	.	15·362 „ „	705
1770,	.	.	15·505 „ „	705
Mean of 3 years,	.		15·459	
1854,	.	.	15·425 feet above O. D. S.	663
1855,	.	.	15·425 „ „	678
1856,	.	.	15·515 „ „	651
Mean of 3 years,	.		15·454	

2d. *Mean Level of Highest Spring-tides throughout the Year.*

1768,	.	.	18·590 feet above O. D. S.
1769,	.	.	18·673 „ „
1770,	.	.	18·816 „ „
Mean of 3 years,	.		18·693
1854,	.	.	19·030 feet above O. D. S.
1855,	.	.	18·873 „ „
1856,	.	.	19·236 „ „
Mean of 3 years,	.		19·016

" The conclusions flowing from a comparison of tidal levels, coincide with those deduced from the measurement of areas, and go to establish the fact which, at first sight, seems at variance with every-day experience, viz., that the sands of the bay in two respects, area and elevation, present evidence of only trifling change, so trifling that from the data given, it would be difficult to pronounce whether they were on the increase or decrease."

Captain Hills has suggested that there may be a regular " cycle of rotation " in the changes that are going on in the Mersey, but time alone can prove whether such be the case. The bar of the Tay seems hitherto to have been less variable. Such changes may be due to certain states of prevailing winds during high tides, and even to the grounding of vessels in the channel. The alteration in the bar of the Tay was attributed by seamen to the loss of a large vessel laden with jute. The effect of such an obstruction as a stranded vessel, in causing the currents to act on the bottom, has been already referred to at page 178, where it was seen that even the bed of the Tay, consisting of heavy gravel, was materially altered in the course of a few tides, and how much greater must be the effect of a vessel's hull swept by the currents of flood and ebb on the soft sand-banks in the estuaries of which we have been speaking. It is well known that vessels grounding on such sand-banks sometimes entirely disappear in the course of a few tides, the opposition they offer to the currents causing a scour which very speedily excavates a hole large enough to bury them out of sight.

[1] Hills's *Hydrography of the Mersey Estuary*, Liverpool, 1858.

Captain Hills mentions two such cases having occurred on the Mersey. Now, were we to suppose a vessel grounding even for a tide on the edge of a bank forming one side of the bar or deep channel of such an estuary, it is quite possible that this, in connexion with some particular state of the winds and tides, might so affect the banks as to give the current, and ultimately the navigable channel, a tendency to shift, which succeeding disturbances might so encourage and increase as ultimately materially to alter the navigable track. Dock and other works are also being formed on the estuary, both of the Mersey and Tay, which, though not decreasing the depth, may possibly so affect the outflowing currents, as, in combination with certain states of wind and tide, to have some effect in varying the courses of the outer channels.

To trace all the movements of the channels and banks of open estuaries to their true origin would indeed be hopeless, for winds and floods, as well as stranded vessels, may each or all have their share in giving a current a slight direction, which, once commenced, may terminate in a new channel and newly formed sand-banks. As an example of the strange freaks, if I may so express it, that are to be met with in the movements of banks, I may refer to a case on the Tay, where a sand-bank, in the course of a single year, changed its position without altering its form. Fig. 58 shows the part of the estuary of the Tay, opposite Balmbreich Castle, where this occurred. In surveying the river in 1833, the bank was found to have the outline and occupy the position shown in dotted lines. In 1834, on resuming the survey, the

T

position of the bank was found to be altered. It had
retained nearly the same outline, but had shifted about

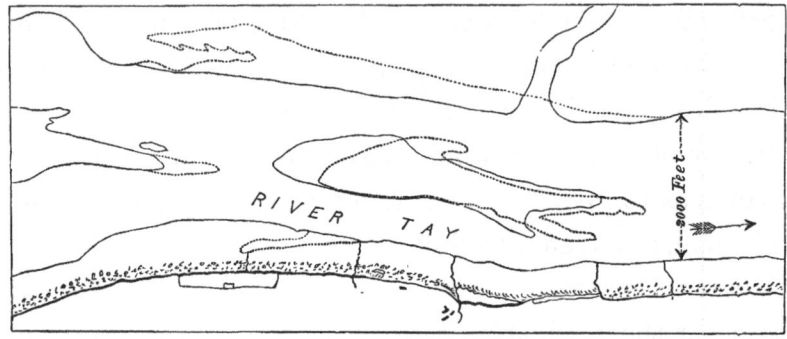

RIVER   TAY

3000 Feet

FIG. 58.

700 feet further up the estuary, and occupied the position
shown in hard lines. I never met with so striking an
instance of what I may term altered *strength* of tidal
currents without alteration of their *direction;* for I
believe the general movement of the particles of sand
composing the bank was caused by decreased power due
to little rain-fall, while nothing had occurred to alter the
direction of the currents, so that the particles of sand
were carried forward in the direction of the flood current,
and deposited so as to present nearly the same outline as
shown in the illustration.

THE WEAR.

Bars of such
rivers as the
Wear and the
Tyne may be
improved by
protecting piers.

Some of our rivers, however, such as the Wear and
the Tyne, have not much of what may be termed the
estuarial features, and require different treatment. The
construction of piers for improving the entrances to
such rivers is often highly beneficial. I shall take as
an example the Wear, which is shown in fig. 59. In

its natural state, such a river as the Wear flows across
the beach from high to low water, in a broad and
shallow channel, the direction of which is ever changing.   It thus forms a long bar or shoal, with broken

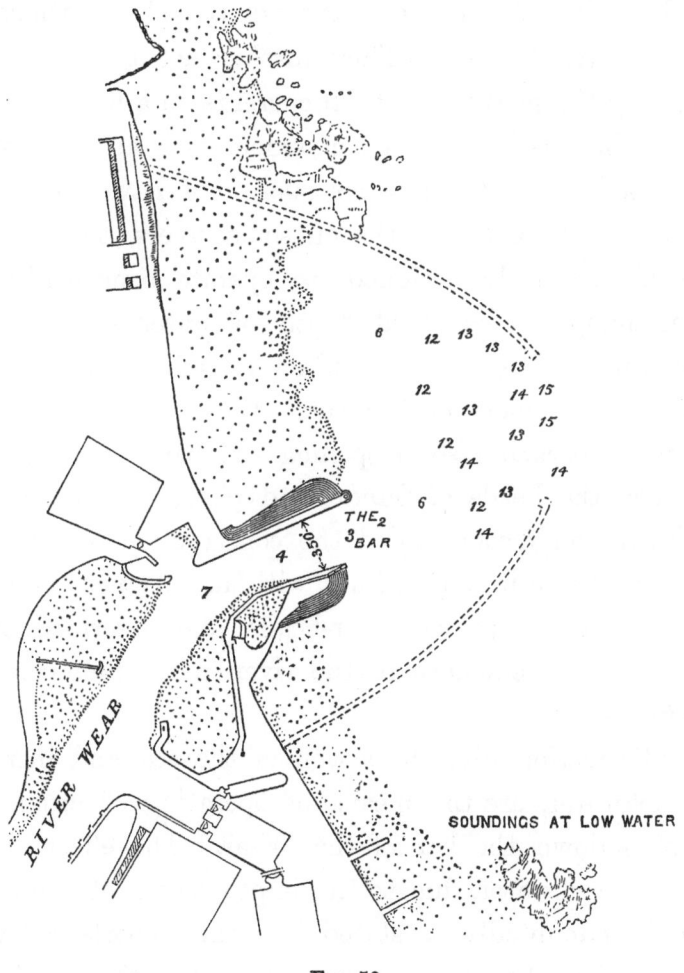

FIG. 59.

water throughout its whole extent.   But the projection of piers across the beach affords shelter from
the waves, and admits of a navigable channel being ex-

cavated and maintained ; and after a vessel entering the
river crosses the short bar, which occurs at or near the
pier-heads, she not only gets into deeper water, but has
the additional advantage arising from the shelter afforded
by the piers. To this extent piers in such situations are
highly advantageous. They further act beneficially in
directing the flow of the tidal currents in a fixed channel
across the beach, and, in connexion with an increase of
tidal capacity in the interior, such as I have mentioned
as the result of the works on some rivers, they cannot
fail, if judiciously designed, to operate beneficially, by
maintaining an increased depth of water on the bar.
Founding on these views, when consulted in 1858 by
the Commissioners of the river Wear, as to the best
means of permanently deepening the bar which extends
between the heads of Sunderland piers, Messrs. D. and
T. Stevenson recommended the construction of covering
piers, as shown in dotted lines, with an entrance 1200 feet
seaward of the present pier-heads, and described their
action on the entrance to the river in the following extracts :—

"Protection from the action of the sea, and increase
of backwater, are the means of operating effectually in
keeping down the bar of the Wear. The effect of increased backwater, due to the improvement of the river,
would undoubtedly, as stated in former reports, act very
beneficially. But nevertheless, so long as the bar is exposed to the undiminished action of the sea during heavy
gales, it must be subject to constant changes in its depth
of water ; and this variableness in the navigable channel,

especially with the larger draught of vessels to be now accommodated, must doubtless be attended with inconvenience and obstruction to the trade of the port, which it is most desirable to avoid. We have therefore considered it necessary to submit to the Commissioners a plan of improvement based on the fundamental principle of protecting the bar from the tendency to heap up or accumulate during heavy seas. . . . We may state generally that the effect of these piers will be to protect the entrance to the harbour, and to allow the tidal scour to act freely on the bottom, and maintain a greater depth on the bar, while the deeper water in which the pier-heads are proposed to be founded, will prevent the bottom at the outer entrance from accumulating or rising, so as to act as an obstruction to such vessels as the interior of the harbour is capable of accommodating."

There are, however, rivers which present very different Bar of the characteristics from those we have been considering, both Mississippi. as regards the fresh-water stream and the action of the sea, and it will be interesting to notice them ; I allude to such rivers as the Mississippi and the Danube.

Mr. Ellet, though founding his views on totally different premises from those I have laid down, also comes to the conclusion that the bars of the Mississippi are not due to the materials deposited by the out-going stream. But I shall give his interesting explanation in his own words. I have, however, no information to enable me to form an opinion as to its correctness. It is based on the fact already described in Chapter V., that at the junction of a river with the sea, the fresh water flows in a stratum above, and distinct from, the salt water, for some distance after entering the ocean.

Founding on this, Mr. Ellet says,—" The velocity of the river is not destroyed, nor very sensibly diminished, at the bars. When the river was rising, but still far from being at full height, I measured the velocity of the current on the bar of the Pass à la Loutre, and found it to vary, at different times and places,- from 3 feet to $3\frac{1}{2}$ feet per second, or from 2 miles to $2\frac{1}{10}$th miles per hour. I measured it also repeatedly on the south-west bar, and found it there 3 feet per second, or about 2 miles per hour. But there are many parts of the river where the speed of the current does not exceed $2\frac{1}{2}$ miles, or even 2 miles per hour, in times of flood, and where it is, notwithstanding, more than 100 feet deep. In fact, on testing the velocity of the south-west pass, 4 miles above the bar, and in 5 fathoms water, I found the current to be but 2 miles per hour,—precisely the same as it was under like circumstances of wind and tide on the bar. The current of the Mississippi sweeps over the bars at the mouths of the passes, and at periods of flood many miles out into the gulf, with a velocity almost undiminished by its contact with the waters of the gulf." " The river water does not mix suddenly with the sea, but rises upon it, floats over it, and rushes far out into the gulf on the top of the dense sea water, by which it is buoyed up. I tested this repeatedly, and found uniformly a column of fresh water, nearly 7 feet deep, in the gulf, entirely outside of the land, and salt water at a depth of 8 feet from the surface, and extending thence to the bottom. The river does not come down with a certain normal depth and speed, and encounter the gulf at the bar. No such process takes place. There is no sudden destruction of velocity, or consequent deposit of suspended silt. But the water of the Mississippi does not move over the surface of the gulf at a speed of 3 feet per second without imparting a portion of its motion to the sea.[1] The fresh water and the salt water take the same direction towards the sea, and with nearly the same velocity, but yet keep separate. This state of things clearly cannot exist at the bottom; for as the river water is for ever coming forward, if the salt water all flowed towards the gulf, it would all be carried out, and river water would take its place. Salt water must come in from some quarter, to supply the current of sea water that is for ever setting towards the gulf, beneath the water discharged by the river. This salt water can only come from the sea, and can only come in along the bottom. It is, in fact, an eddy that is here at work, the movements being in a vertical instead of a horizontal plane. Now, the question is, How does this account for the existence of the bar? The fresh water running out cannot produce deposit, for it has velocity enough to sweep away a foundation of coarse gravel. The outpouring salt water immediately beneath the fresh, cannot produce deposit, because it also has a velocity seaward strong enough to remove anything that is brought down the Mississippi. The salt water that is coming in might produce, and I

[1] This is in harmony with Venturi's well-known experiments, from which he found, that a body of water in motion leads or drags with it the particles of water at rest with which it may be in contact.

doubt not does produce, a deposit, for it passes over the soft muddy bottom of the gulf, and moves into the river, and along the bar, at a very slow rate. According to these facts, and this reasoning, there must be usually on the bar three distinct strata : 1*st*, Fresh water, running out at top, found by experiment on the s.w. bar to have a velocity of 3 feet per second. 2*d*, Salt water below the fresh, also running out with nearly the same velocity as at top ; and 3*d*, Salt water coming in slowly along the bottom, and apparently a sheet of salt water between that running out and that coming in, which will be without motion.

"But as already said, and as is obvious, all the sea water that comes in must go out again. It comes in along the bottom, and it must go out between the column of salt water coming in and that of the fresh water going out. Each particle of salt water, therefore, must change its direction and position in elevation. It must pass from an inward-bound lower stratum to an outward-bound upper stratum. But in passing through this change of motion, its velocity up stream must be neutralized. It passes, to use a technical term, the *dead point*. At this point it may cease to bear its whole burden of mud, which it has brought from the gulf further forward. It leaves it, or a portion of it, at the turning-point. This turning-point is the place where the bar for the time being is in process of formation. But as the upper and lower strata are moving in opposite directions, the intermediate column must of necessity have a rotatory motion. That motion must be shared by the lower column of salt

water, and this turning-point must therefore be formed
at the same time at different places along the bar."

The Danube is interesting as an example of a large
river having been successfully treated by the construction of piers, and also, as the reader will find, in the origin
of its "bar" as distinguished from the sea-bars treated of
in the beginning of this chapter.

Bar of the Danube, its cause, and works for its improvement.

In 1856 the "European Commission of the Danube"
was appointed under the Treaty of Paris, and consisted
of seven delegates, representing England, Austria, France,
Prussia, Russia, Sardinia, and Turkey, and its object was
to improve the bar of the river, and open the navigation
to the traffic of all nations. The Danube, after flowing
over a course of 1700 miles, and draining 300,000 square
miles of country, enters the Black Sea by three separate
mouths—the northern called the Kilia, the central the
Sulina, and the southern the St. George's mouth. The
first duty of the Commission, with the advice of Sir
Charles Hartley, who was appointed their engineer, was
to select one of the three mouths for improvement, which
was by no means an easy task, as each of them presented
advantages peculiar to itself, and after much consideration
the Sulina or central channel was selected, and although
considerable difference of opinion existed as to the propriety of the choice the result has shown that the course
adopted was judicious.

The Danube discharges in ordinary flood no less than
twenty millions of cubic feet of water per minute, enters
a tideless sea, and we have a totally different class of
phenomena to deal with from those of the kind which I

have just been considering. The river brings down an amount of detritus which has been ascertained by Sir Charles Hartley to be equal to 27 cubic inches per cubic yard, and to be equal, in cases of high flood, to no less than 600,000 cubic yards of solid deposit in 24 hours. Like the Mississippi and the Nile, the Danube owes its extensive delta to the gradual accretion of this sedimentary deposit, and the bar at its mouth is due to the same action. It therefore differs entirely from the bars in this country, as is well exemplified in the fact, as has been already stated, that, whereas in our harbours the bars are always *deepest* when the sea is calm and the rivers are *in flood*, and therefore most efficient as scouring agents, at the Danube the bar is, on the contrary, invariably *shallowest* when the river is in flood, because it is then charged with a larger amount of detritus.

Another feature of difference in the treatment of such a case as the Danube is to be found in the circumstance that there is no reversal of the current due to tidal influence, and therefore it is unnecessary, in fixing the direction of the piers, or indeed in designing any of the works, to provide for the admission of tidal water to act as a scour on its return to the ocean, a provision which always demands special attention in designing tidal works on our coasts.

The works executed at the Sulina mouth, as shown in fig. 60, consist of a north pier 4640 feet in length, and a south pier 3000 feet in length, both built of *pierres perdues* surmounted by a timber staging, with an entrance between of 600 feet, and the slightness of their structure indicates the modified character of the waves to which they are exposed.

The works, which are understood to have cost about £100,000, are highly creditable to the talent and energy

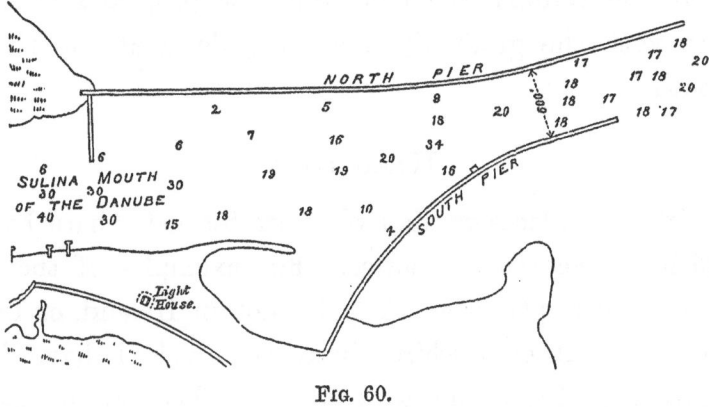

FIG. 60.

of Sir Charles Hartley, and have now been completed for nine years, and their effect has been most satisfactory, as proved by the fact that, previous to their construction, the depth on the bar never exceeded 11 feet, and frequently fell to 8 feet; whereas, according to the last accounts from Sir Charles Hartley, the depth for the last five years has never been less than 15 feet, and has often been as much as 17½ feet.

It is obvious, however, that as the Danube must continue to bring down an enormous mass of detritus, so, in course of time the works which have proved so successful must be extended—an event which has been fully anticipated by its projectors, and in this respect we find an interesting difference between such works as the Danube piers and the harbour works of this country, for here the object being to prevent the waves from acting on the bottom, the engineer extends his works out into a depth of water where there is little or no disturbance of the bottom, and if this is once secured he may calculate on the increased depth of water remaining permanent, whereas at the Danube the piers must be projected to keep pace with the gradually increasing delta at the river's mouth.

HARD BARS.

In other places we find what are termed "hard bars," which I have still to notice. For examples of these I refer to such places as Ballyshannon in Ireland, or Loch Fleet in Sutherlandshire, both of which I have had occasion professionally to examine. The bar at Loch Fleet, for example, is composed of boulders firmly imbedded in a mass of indurated gravel, and is obviously a continuation of a bed of similar formation which seems to traverse the coast at that place, while that at Ballyshannon is simply a heap of large boulders. The consequence, in either case, is that no scouring power can make the least impression on the channel. Such bars are entirely due to the hardness of the bottom, and though their hardness makes such obstructions troublesome to remove, and though, moreover, they are generally in exposed situations, still they are comparatively easily treated by the engineer, and an encouraging prospect is always held out that their removal will be attended with *permanent* benefit, since by excavating a channel through them we at the same time remove the evil and its cause.

Groynes.

The entrances to some rivers are greatly impeded

by shingle or gravel, carried along by the waves from the adjoining shores, and deposited in the channel. I have known such deposits, if not wholly removed, at least greatly modified by erecting groynes across the beach in such a position (depending on the direction of the heaviest seas), as either to collect the shingle and retain it until it can be carted away, or to lead it past the harbour mouth altogether, and force it onwards to a place of deposit in an adjoining part of the coast; and when this can be brought about, the engineer may congratulate himself on having designed a very successful work.

CHAPTER XIII.

RECLAMATION AND PROTECTION OF LAND.

Schemes for gaining land and improving navigation not generally compatible—Illustrated by the Dee—Depression of low-water line apt to mislead, as tested in the Lune—Increase of tidal water at the Lune, Tay, and Ribble, and its effect as a scouring agent—Adjoining property benefited by river improvements—Process of land-making depends on amount of matters held in suspension—Heaviest matters found next the sea in tidal estuaries, the reverse in such rivers as the Danube, etc.—Size of particles which estuaries are capable of carrying—Weight of different deposits in the bed of the Clyde—Quantity of matter held in suspension by different rivers—Formation of deltas—Level of vegetation in marsh lands—Works for protection of marsh lands—Works for protection of land in open estuaries.

Schemes for gaining land and improving navigation not generally compatible.

SUCH twofold schemes as have for their ostensible object the improvement of rivers and the *formation of land*, have generally been unsuccessful in benefiting navigation. I do not affirm that river works, constructed on the principle that has been advocated in the foregoing pages, have not the effect of *making land*, in the particular sense in which I shall afterwards explain it; but *land-making* is no part of sound *River Engineering*. Judiciously designed works may, as I propose now to show, reclaim and protect land, while at the same time, as their *primary* object, they improve navigation; but I know of no case where the interests of navigation have been promoted by any measure which has for its main object the construction of walls designed to convert large tracts of tide-covered sands into cultivated fields.

RIVER DEE.

I shall refer to the Dee, in Cheshire, as an aggravated instance of the incompatibility of the two interests.[1] The outline of this river is shown in Plate V., from a survey by Messrs. Stevenson, made in 1838. The River Dee Company, incorporated by Act of Parliament in 1732, have from time to time reclaimed from the upper part of the estuary a large tract of land, extending to about four thousand acres, which is now in full cultivation; and alongside of this gradually gained territory the river has been conducted from Chester to near Flint, in a narrow canal of about 8 miles in length, and 400 feet in width. A considerable portion of land has also been reclaimed on the Flintshire side of the estuary, though not by the proprietors of the Dee Company; and it is believed that the aggregate amount which has from first to last been gained from the sea is about seven thousand acres. Now, it is well authenticated that previous to the commencement of the land-making operations on that river, there was a depth of not less than a fathom at low water of spring-tides up as far as Burtonhead, and that there was an anchorage for vessels of the largest size opposite to Park-gate, the positions of which places are marked on the plan. But when I surveyed the Dee in 1838, the depth of 6 feet was not found for more than six miles below Burtonhead, the low-water features of the estuary having

---

[1] *Great Britain Coasting Pilot*, by Captain Greenville Collins, Hydrographer in Ordinary to the King's Most Excellent Majesty, London, 1767 ; *Reports to the Admiralty*, by Captain Washington ; *Report of Tidal Harbour Commissioners ; Report* by Messrs. Stevenson, 1839.

been forced to that extent farther seawards by the extensive reclamation of land in the upper part of the estuary, and the consequent diminution of the tidal scour. It cannot, we think, be disputed, that the effect of the works executed on the river Dee, whatever may have been the anticipations of their projectors, has been to shut out the sea, and form land at the expense of the navigation.

The process followed in carrying out the land-making works was to construct a high bank, rising 9 feet above the level of high water, so as to confine the river to the south side of the estuary. The tidal water, which was admitted to flow freely between the bank and the north coast, quickly deposited layer after layer of sand and silt, and in fact shut itself out, and so soon as the surface had attained a sufficiently high level, a cross bank was constructed between the main embankment and the north shore, and thus the large area shown on the plan was *bit* by *bit* reclaimed. The reclaiming banks were gradually strengthened and pitched on the outer face, and substantial sluices were formed, which are shut against the ingress of the rising tide, but being open at low water, allow the drainage-water to escape from the reclaimed ground, some of which is still below the level of high water.

It will at once be seen that the works constructed in the process of land-making, as carried out on the Dee, were totally different from the low-water training walls which I have described. The system pursued is in fact in direct opposition to the principles of River Engineering which have been laid down, and the result, as has been seen,

has not been favourable to navigation, at all events in
the case of the Dee.

But an objection has sometimes been raised to navi- Depression of low-water line apt to mislead.
gation works formed on the principles laid down, which it
is necessary to notice.   The change produced on the
relative levels of the low water and the banks, by the
depression of the low-water line, described at page 250,
have sometimes led to considerable misapprehension.
This lowering of the surface of the water when the river
is confined by walls in the lower part of an estuary, invariably conveys the impression that a great rise has
taken place in the level of the adjoining sand-banks, and
it has consequently been thought that the erection of
river walls is inconsistent with the principle of non-
exclusion of tide-water which I have been advocating.
But leaving out of view, for the present, the enormous
gain to the navigation by the increased scour, due to the
enlargement of the channel, as explained at page 224, it
can be shown that even the increase of the sand-banks
may be greatly misunderstood and exaggerated.

In its natural state, the channel of such an estuary as
the Lune or the Ribble, as already explained, is subject
to constant change of position.   I have seen many acres
of marsh or grass land in such estuaries carried off by the
sea, and the solid matter of which they were composed
scattered over the shores and sand-banks.   Now, the
effect of fixing the channel by means of walls, in the
manner recommended, is to form one permanent navigable
track ; and the banks on either side, being no longer subject to the periodical inroads of the river or tides, gradu-

U

ally rise in elevation until they are capable of producing vegetation, and ultimately become what are termed marsh lands. When a river channel has been thus fixed and confined by walls, I have ascertained by repeated observation that the tidal water comes up the channel in a comparatively *pure state*, instead of being loaded with particles abraded from the sand-banks and marshes. It has also been found that the process of deposit at the sides of an estuary so improved goes on very slowly after it has reached a certain stage ; for the materials deposited on the upper parts of the banks are, as afterwards more particularly described, exceedingly fine, and are carried only by the highest tides, which seldom reach those elevated portions of the shores. From all these considerations I infer that the effect of river-walls upon an estuary is mainly to prevent the constant disturbance of the materials of which the banks are composed, but not necessarily to occasion additional accumulations.

As tested in the Lune.

I had an opportunity, at the Lune, of testing by actual measurement in how far the raising of the banks, caused by the erection of the walls, was due merely to a new disposition of the materials which originally filled the bed of the estuary, or to additional foreign matters deposited in consequence of the operations. I am not aware that similar observations with this object have been made on so large a scale ; and as they are highly important in assisting our views as to the economy of tidal estuaries, I shall give a brief notice of them.

Increase of tidal water at the Lune and Tay.

On referring to the chart of the Lune, Plate VIII., the changing nature of the channel will be seen from the

different courses in which it flowed, as shown by dotted
lines.   To obviate this, training walls and other works
were constructed, which caused, as might have been expected, a very considerable alteration in the position and
form of the sand-banks in the estuary.   This alteration,
in connexion with the depression of from two to three
feet in the low-water level of the river, was apt to lead a
casual observer to suppose that a great accumulation of
sand had taken place, and consequently that a corresponding amount of backwater had been excluded, and observations were made to determine the state of the case.
Fig. 61 represents the changes that were produced by the
works.   Over the whole area, which is represented as
covered by sand, a deposit had taken place, the banks
being higher than formerly, whereas the whole area included in *hatched* lines had been scoured, the banks
having been lowered.   A careful calculation was made,
founded on numerous sections taken in 1838 before the
works were commenced, and in 1851 after their completion.   The result of this investigation was, that after the
completion of the works the amount of deposit on the space
shown as sand in the cut was 3,070,146 cubic yards ;
while the amount of scour on the space shown by hatched
lines was 2,810,449 cubic yards ; giving an excess of deposit of 259,697 cubic yards.   But the amount stated as
having been scoured does not include what has been taken
away below Glasson and Basil points ; which has doubtless been deposited in the bank above.   The survey of
1838 did not afford data for ascertaining the amount
of what had been scoured below Glasson with sufficient

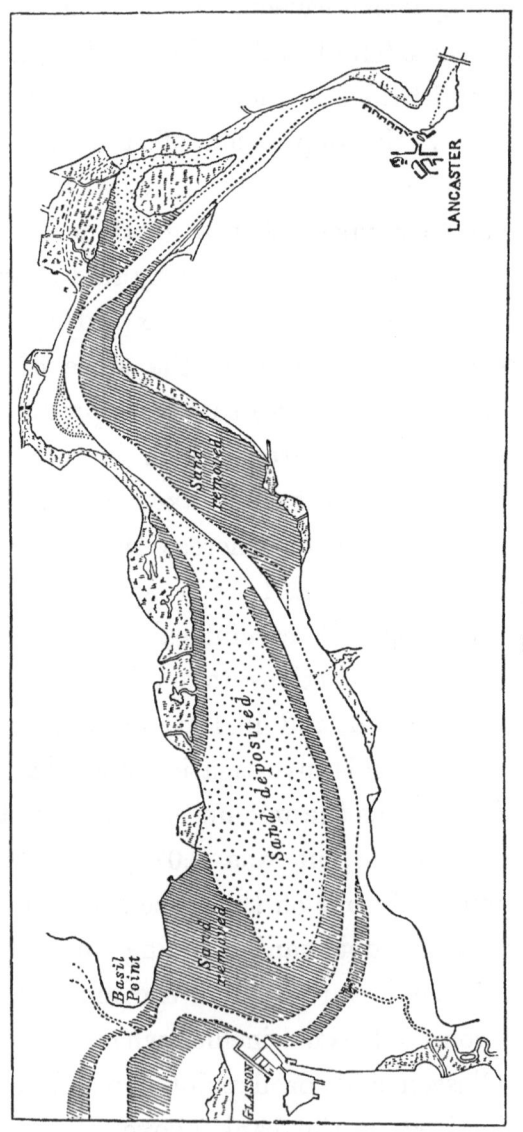

FIG. 61.

accuracy to admit of its being included in the foregoing
calculations. But an amount of scouring was ascertained
to have actually occurred at that place, which was amply
sufficient to counterbalance the surplus of 259,697 cubic
yards of deposit, as given in the above statement.

Such a result may indeed be expected ; for it is difficult to conceive in what way parallel walls formed in an
estuary can operate either in bringing down additional
alluvial matters from the river above, or in bringing up
additional detritus from without the bar.

Holding these views, and supported by the actual
observations made in the case of the Lune, I, therefore,
conclude that the tendency of works executed in accordance with the principles laid down *is not necessarily to
produce additional accumulation of matter, but simply to
alter the disposition of the existing materials of which the
bed of the estuary was originally composed.*

But assuming that in some cases a deposit does take
place, and that the gradual rising and ultimate reclamation of marsh land excludes a certain portion of tidal
water, it is important to consider in how far such
abstraction of water is counterbalanced by the navigation works, and on a full consideration of the matter
it will be found that the compensation afforded by well-
designed works is very much greater than is generally
supposed.

I have already said, at page 226, in considering the
question of scouring power, that the aggregate annual
effect of the additional water gained by the operations
on the Tay was equal to two months' flow of the river

in its ordinary state. I have also shown, at page 283, that the water so gained acts on the low-water channel, and is therefore calculated to produce what may be called the maximum scouring power. As we are now speaking of land reclamation, it may be well still further to consider what relation this additional scouring water bears to the sheets of shallow water which are spread over extensive areas when covered by high tides. Perhaps I shall best give an idea of this by stating one or two examples from actual practice.

As regards the Tay, we have seen that the *additional* quantity of water filled and emptied every tide is one million cubic yards, and this occurs 730 times in the year. Now, it is interesting to ascertain in such a case what area of land could be enclosed without impairing the beneficial effect of the tidal scour. Assuming that the marsh lands proposed to be enclosed may have been covered five or six times during each set of spring-tides, or say 144 times during the year, to the *average* depth of 1 foot, the effect produced upon the navigation, by the *acquired* and *abstracted* tidal water, may be expressed in the following manner :—

Founding on the formula, $S \propto V T$, already given at p. 283,

Cubic Yards.

Let VT = the acquired water, viz., $1,000,000 \times 730 = 730,000,000$

and vt the abstracted water per $\Big\}$ acre of reclaimed land, $= \dfrac{43,560 \times 1}{27} \times 144 = 232,272$

then $\dfrac{VT}{vt} = \dfrac{730,000,000}{232,272} = 3143$ acres

—thus showing, that the improvements effected on the

river were such as to allow of the reclamation of 3143 acres of land at the level referred to, without diminishing the original scouring effect of the tidal water. In other words, on the Tay, there might be enclosed an area of marsh land covered to the *average* depth of 1 foot during spring-tides, equal in extent to the whole surface of the river from Perth down to about 2 miles below Newburgh, before excluding an amount of water equal to the aggregate quantity brought in by the navigation works.

On the Ribble, also, additional tidal water, amounting to 1,745,000 cubic yards, has been gained by the lowering of the river, and, applying the same calculations, this represents an extent of marsh land of 5484 acres, being equal to the whole area of the estuary from Preston to about a quarter of a mile below the Naze Point. It will thus be seen that, even if we assume the land to be covered to a greater average depth than 1 foot, there is ample room for reclamation, within certain limits, on *properly treated tidal estuaries*, with advantage both to the interests of the navigation company and the proprietors of land.

It is obviously highly important if the two objects of *river* and *land* improvement can be carried on simultaneously; and to a large extent this, as has been shown, is perfectly practicable. The attempts of proprietors to protect the foreshores of their lands from the encroachments of rivers in tidal estuaries, are often attended with great expense; and if those efforts prove for some time effectual in warding off the approach of the channel, the land speedily takes on vegetation, and is fit for

Adjoining property benefited by river improvements.

pasture. But the tenure by which such property is held is very slight; and the spot which to-day affords grazing for cattle may in a few tides become the navigable channel of the river. Now it is obvious that the perfect protection from such encroachments afforded by the training and guiding of the low channel by longitudinal walls, adds materially to the value of the adjoining property; for not only is the land beyond high-water mark completely protected from encroachment, but the marsh lands bordering the estuary become, in fact, *permanent property*, and not an ever-changing benefit, held for one year and probably lost the next. Marsh lands so protected from waste are still, it is true, liable to be flooded by high tides; a circumstance, however, which is considered by some persons not injurious, but rather beneficial for marsh pasture.

Process of land-making.
The process of land reclamation to which I have alluded is generally termed "warping." In most cases the tide is permitted to flow freely over the surface, and whatever is deposited at slack tide contributes to the accretion. Sometimes the land-making is hastened by forming banks with sluices, and retaining the water till it deposits the whole of the matter in suspension, and then permitting it to run off slowly.

It is obvious that the rate at which the process goes on depends on the quantity of matter held in suspension, which varies in different estuaries. The size of detrital particles which are carried by the currents of estuaries depends on the velocity of the stream, the nature of the bottom along which the detritus is moved, as well as the shape of the particles of which the detritus itself is composed, and is altogether a subject so dependent on special circumstances, that it is impossible to lay down rules which can be generally applicable. I must therefore content myself with giving results as communicated by different authorities. Before doing so, however, I may state a rule which I have found to apply to the Dee, Ribble, Lune, and Wear, and which I believe to be generally applicable. It is, that in all tidal estuaries *the heavier sands and deposits are found on the banks at the mouth of the estuary, and the particles are lighter as we recede inwards.* I have tested this on the rivers above mentioned, and others, by agitating equal quantities of sand and deposit (taken from different parts of the tidal estuary) in equal quantities of water, and observing the time which elapsed, in each case, before the materials were deposited and the water assumed a state of purity.

Heaviest matters in tidal estuaries found next the sea.

The result of these observations proved that the sand of outer or seaward banks, where the currents were strong, was composed of large particles, held in suspension only a few seconds, while in the inner parts of the estuary the deposit decreased in weight, and *generally* that it decreased from low to high water where the currents were weak and where the silt was exceedingly fine, and remained in suspension, in some cases, even for hours after the agitation of the water.

It will be seen that the rule I have stated is at variance with that propounded by Frisi, and also by Sir H. de la Beche,[1] in the following terms :—" Where the velo-

[1] De la Beche's *Geological Manual.*

city of a river is sufficient to produce attrition of the substances which it has either torn up, collected by undermining its banks, or which have fallen into it, they gradually become more easy of transport, and would, if the force of the current continued always the same, be forced forward until the river delivered itself into the sea; but as the velocity of a current greatly depends on the fall of the river, the transport is regulated by the inclination of the river's bed. Now it is well known that this inclination varies materially even in the same river, so that it may be able to carry detritus to one situation, but may be unable to transport it further, under ordinary circumstances, in consequence of diminished velocity. As a general fact, it may be fairly stated that rivers, where their courses are short and rapid, bear down pebbles to the seas near them, as in the case of the Maritime Alps, etc.; but that where their courses are long, and change from rapid to slow, they deposit the pebbles where the force of the stream diminishes, and finally transport mere sand or mud to their mouths, as is the case with the Rhone, Po, Danube, Ganges, etc." This holds true in the case of such rivers as those to which Sir H. de la Beche refers; but it will be found, as I have stated, that the case is exactly reversed in tidal estuaries.

Size of particles which streams are capable of carrying.

The following are the results of experiments made by Bossut, Du Buat, and others, on the size of detrital particles which streams flowing with different velocities are said to be capable of carrying :—

3 in. per sec. = 0·170   mile per hour will, just begin to work on
fine clay.

6 „   „   = 0·340   do., will lift fine sand.

8 „   „   = 0·4545 do., will lift sand as course as linseed.

12 „   „   = 0·6819 do., will sweep along fine gravel.

24 „   „   = 1·3638 do., will roll along rounded pebbles 1 inch
in diameter.

3 ft.   „   = 2·045   do., will sweep along slippery angular stones
of the size of an egg.

The following experiments were made by Mr. T.
Login, C.E., and are given in the *Proceedings of the
Royal Society of Edinburgh*, vol. iii. p. 475. They were
made with a stream seldom exceeding half an inch in
depth; and are as follows :—

Nature of Materials.	Rate of sink-ing in water.	Current required to move.	
	Feet per minute.	Feet per minute.	Mile per hour.
Brick-clay when mixed with water, and allowed to settle for half-an-hour, ...	·566	15	·170
Fresh-water sand, ...	10	40	·454
Sea-sand, ...	11·707	66·22	·752
Rounded pebbles about the size of peas, ...	60	120	1·37
Vegetable soil, ...	...	50	·56

Brick-clay in its natural state was not moved by a current of 128
feet per minute, or 1·45 mile per hour.

The following statement by Mr. William Bald, of
experiments made on materials taken from different parts
of the bed of the Clyde, shows the variety of materials
found in the same stream, and is a valuable record of
the weight of the deposits which form the beds of our
tidal rivers :[1]—

Weight of de-posits in the bed of the Clyde.

[1] *Minutes of Proceedings of Institution of Civil Engineers*, vol. v. p. 330.

Deposits.	Lbs. to cubic feet.	No. of cubic feet to the ton.
Fine sand and a few pebbles laid in the box loose, not pressed, nearly dry, . .	87	26
Do. do. pressed, ...	92	24
Mud at White Inch, dry, and firmly packed; contained very fine sand and mica, .	97	23
Wet mud, rather compact and firm, well pressed into the box, . . .	115	19
Wet, fine sharp gravel, well pressed, . .	124	18
Wet running mud, ·.	122½	18·1
Sharp dry sand deposit in harbour, . .	92	24·3
Pt.-Glasgow Bank, (sand) wet, pressed into a box,	120½	18·6
Sand opposite Erskine House, wet, pressed, .	116	19·3
Alluvial earth, pressed, ...	93	24
„ „ loose, ...	67	33

I found the gravel of the Tay to be 18 feet to the ton.

Quantity of matters held in suspension by different rivers.

The *quantity* of solid matter carried or held in suspension by rivers, has also been made the subject of observation. Different observers whose remarks have come under my notice, have stated their results in different ways, some giving the *weight* and others the *bulk* of detritus. Thus Mr. Ellet says that the sedimentary matter transported by the Mississippi forms $\frac{1}{3000}$th part of the *volume* discharged by the river.[1] Mr. T. Login, C.E., Pegu, states, in a paper on the Delta of the Irrawaddy, read before the *Royal Society of Edinburgh*, session 1857, that the waters of the Irrawaddy contained $\frac{1}{1700}$th part of their *weight* of sediment during floods, and $\frac{1}{5725}$th part of their *weight* when the river was in a low state, and gives the mean deposit at 8 inches per cubic yard. Mr. Leonard Horner found that the water of the Rhine at Bonn contained from $\frac{1}{12500}$th part of its

[1] Ellet, *On the Ohio and Mississippi.*

weight during floods to $\frac{1}{20734}$th part of its weight in a low state.[1] Captain Denham found that the tidal water of the Mersey contained 29 cubic inches of solid matter in every cubic yard during flood-tide, and 33 cubic inches in every cubic yard during ebb-tide.[2] Sir Charles Lyell says:—"Hartsaeker computed the Rhine to contain, when most flooded, 1 part in 100 of mud in suspension. By several observations of Sir George Staunton, it appeared that the water of the Yellow River in China contained earthy matter in the proportion of 1 to 200. Manfredi, the celebrated Italian hydrographer, conceived the average proportion of sediment in all running water to be $\frac{1}{175}$th. Some writers, on the contrary, as De Maillet, have declared the most turbid waters to contain far less sediment than any of the above estimates would import; and there is so much contradiction and inconsistency in the facts and speculations hitherto promulgated on the subject, that we must wait for additional experiments before we can form any opinion on the subject."[3]

But assuming 18 cubic feet of solid matter to weigh a ton, the following table presents a fair view of the cubic measure of solid matter, and the ratios of volume and weight in each case. In submitting this table, I must observe that the discrepancies in the statements are so great, that further observations are necessary before any satisfactory conclusion can be arrived at; but I give the

[1] *Arcana of Science and Art*, 1835.

[2] *Observations on the Mersey*, by Captain H. M. Denham, R.N., Liverpool, 1840.

[3] *Principles of Geology*, by Charles Lyell, F.R.S., London, 1830, vol. i. p. 247.

results as they have been stated by their respective authorities :—

Name of River.	Cubic inches of solid matter in every cubic yard of water.	Ratios of volume of solid matter to volume of water.	Ratios of weight of solid matter to weight of water.
Mississippi, mean, . .	15·5	$\frac{1}{3000}$	$\frac{1}{1511}$
Irrawaddy, in flood, . .	11·71	$\frac{1}{3984}$	$\frac{1}{1700}$
Do., ordinary state, .	4·1	$\frac{1}{11370}$	$\frac{1}{5725}$
Rhine, in flood, . .	1·87	$\frac{1}{24949}$	$\frac{1}{12500}$
Do., ordinary state, .	1·13	$\frac{1}{41288}$	$\frac{1}{20734}$
Do., mean, . . .	1·5	$\frac{1}{31104}$	$\frac{1}{15698}$
Mersey, flood-tide, . .	29·	$\frac{1}{1609}$	$\frac{1}{817}$ } salt
Do., ebb-tide, . .	33·	$\frac{1}{1413}$	$\frac{1}{728}$ } water.

From this table it appears that the Rhine, as compared to the others, is exceedingly pure ; while the waters of the Mersey, on the other hand, hold in suspension a very large amount. It must be kept in view, however, that the source whence the sedimentary matter in the Mersey is derived, is very different from any of the other cases mentioned in the table. The main part of the solid matter in suspension in the Mersey, and indeed in all our tidal rivers, is sand, stirred up by the flowing tide, and deposited again during the ebb-tide. The sedimentary matters in such rivers as the Mississippi or the Irrawaddy, on the other hand, are borne down from the low tracts of alluvial country through which it flows, and form a constant and consequently increasing deposit at the lower parts of the river.

Formation of deltas.

In all cases where the tidal currents across the mouths of such rivers are languid or altogether absent, as in the

Mississippi, the Nile, the Danube, and other continental rivers, the deposits brought down are not carried away, but form deltas, which collect with greater or less rapidity in proportion to the quantity of material brought down and the depth of water in which it is deposited. Mr. Ellet computes the delta of the Mississippi at 40,000 square miles in extent, its average length from north to south being 500 miles. Assuming the sedimentary matter brought down at $\frac{1}{3000}$th of the volume of water, and the discharge of the river at 21,000,000,000,000 cubic feet per annum, he estimates that this vast accretion of deposited stuff must have formed at an average rate of 1 mile in 99 years, giving a period for its entire formation of something like 45,000 years! Sir H. de la Beche has, however, with reason suggested that deltas would increase most rapidly at the first period of their formation, on account of the greater declivity of the river, and the supposition that the detritus from the interior would become gradually less, from the equalization of levels and the fewer asperities that agents have to act on ; and thus it seems impossible to calculate from the present rate of accretion the time which the whole mass has taken to accumulate.

The depth of deposit annually left on the shores of estuaries varies as much as the amount held in suspension by the waters. M. Bouniceau[1] states that marshes on the Seine require twelve years to rise to the level of high water, and gives thirty years for a similar action to take place on the Bay of Vays, and eighty years on the

---

[1] *Constructions à la Mer*, par M. Bouniceau, p. 185.

Scheldt. Similar variations are to be found in statements made by various authors.

The most rapid deposit which has come under my
notice was near the mouth of the Avon, at the Severn,
where the channel between Dumball Island and the shore
was silted up to the extent of 32 feet in 7 years. Fig. 62
represents a section of a stream where the summer water
channel was deepened and confined by longitudinal walings
for sanitary purposes. The river when in flood raised the
banks on either side about 3 feet in seventeen years, the
stuff being deposited in regular layers of sand and silt.
But after the banks attained the height represented in the
cut, the floods began to act on the sides of the channel,
and the stream is now wasting away the accumulations

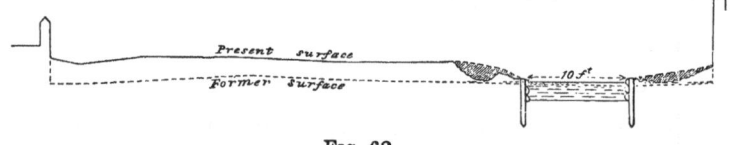

FIG. 62.

that have been gradually made, and this wasting action
will no doubt go on until the sectional area is large enough
to allow the floods to pass off without a velocity sufficient
to carry away the banks. The dotted line represents
the original channel, the hard line the deposited banks,
and the hatched portion what has been wasted away.
This tendency to enlargement close to the edge of the
stream corresponds to what is represented as having
taken place at the Lune, at p. 308, where it will be observed a narrow strip of scouring is shown on either bank.
The interval opposite to the words "sand deposited,"

where there is no hatching, is hard ground, and no sand ever lay on it.

The cost of reclaiming land covered to a considerable depth by the tide is very great.   The estimate for enclosing 10 acres of land for an extensive and important public work on the shore of an estuary where the rise of tide was 10 or 12 feet was £3000 per acre, and on the same estimate, but in a situation not so exposed, the cost of enclosing 30 acres for dock purposes was £2000 per acre.   It is, however, to the reclamation by means of low banks of sheltered marshes on a high level that my observations must refer; and so much depends on the situation as well as on the area enclosed by a given length of bank, that no idea of the cost of such works can be given that can prove generally applicable.   It is, indeed, for the same reason that I have avoided throughout the whole of this treatise giving the cost of the works I have described, as a certain expenditure in one situation might effect either far more or far less work if laid out in a different locality, and in no department of engineering does this hold more true than in river and marine works.

The process of reclamation in all cases goes on very slowly after it has reached a certain stage, because as the banks rise they are more seldom covered by the tide, and the materials deposited on the inner and higher parts of the banks are, as already stated, exceedingly fine, and are carried only by the highest tides, which seldom reach them.   Mr. Park has found on the Ribble the first indications of vegetation to appear about the level of high water of neap-tides, and this corresponds with my

Level of vegetation on marsh lands.

x

own observations at other places.   Mr. Gordon[1] also found that in the Norfolk estuary " the samphire began to settle on the sands which the neap-tides just cover," and that " grass began to grow about one foot above the samphire level," so that the level stated may safely be taken as that at which vegetation commences on the estuaries of this country.   The surface will gradually rise by succeeding deposits, till at last it reaches nearly the limit of high-water spring-tides, which I have found to be the height of different marsh lands.   Mr. Mitchell has found the same result in the United States, as on com-paring the height of different marsh lands their level corresponded to that of " the ordinary high-water level."[2]

Fig. 63 is a section illustrating the manner in which such marsh lands are formed.   The upper portion nearly at the level of high water is what is called " marsh," or " outmarsh," and is fit for grazing.   In some places it is covered with reeds.   Below this level to half tide the surface is covered with occasional patches of samphire ; farther down there is what is called " slob," consisting of sand covered with mud ; and lower down there is sand, more or less pure according to the situation.

Works for pro-tection of marsh lands.

Such marsh lands as those I have described, if left unprotected, must remain for ever liable to be covered during high floods or tides, and therefore cannot be said to be available as arable land without the erection of considerable works for the purpose of protecting them from floods and providing for their effectual drainage.

[1] Report on Norfolk Estuary, by L. D. B. Gordon, C.E., Glasgow, 1856.

[2] On the Reclamation of Tide Lands, 1869.

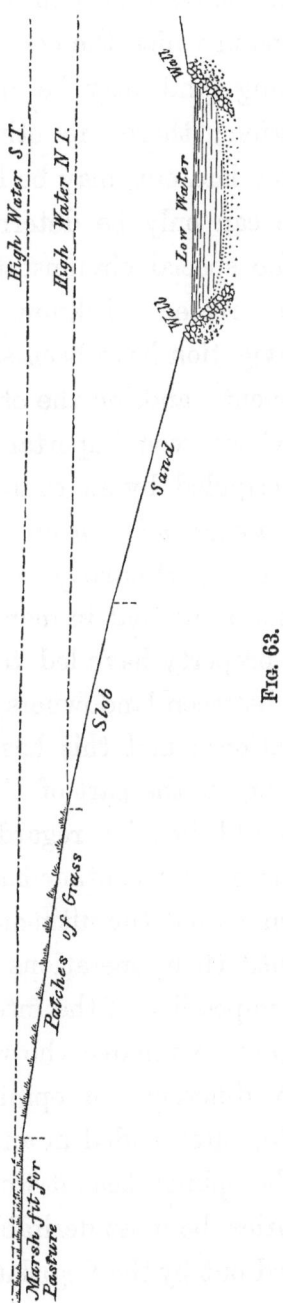

Fig. 63.

The erection of all such works should be well considered. There are situations in which the construction of embankments for protecting land may be injurious to the interests of navigation; there are others in which such works, if judiciously laid out, may be harmless; but their effect in any case can only be determined by a careful consideration of the special circumstances of the locality in which they are erected. I know many cases where the interests of navigation have been sacrificed by unwarrantable encroachment; and, on the other hand, instances are not wanting where even important works have been embarrassed and crippled by an over-cautious regard to the principle of *non-encroachment* on the high-water line. With reference more particularly to the operations of landowners, it is notorious that in many cases attempts to reclaim or protect property have led to serious and costly legal proceedings between landowners and the local conservators of navigations; and this has in some instances arisen from a feeling, on the part of the landowners, that their operations could not be regarded as prejudicial. The local conservators, on the other hand, have generally no means of knowing what the ultimate intentions of the landowners are until their operations have proceeded so far as to render it impossible, if the interests of navigation require it, to stop or to remove the works without considerable loss. A difference of opinion has thus been raised, which has too often ended in an expensive lawsuit. I have long held the opinion that it would in many, if not in all, of our estuaries, be most desirable to have a line of conservation marked out by the Legislature for the regulation of all works for the protection of land, just as we now have lines defining the boundaries of sea and river fishings. Were such a line of conservation defined, the landowners could then with confidence, and without risk of challenge, enter on such works within the legalized boundary as they considered necessary for the protection of their property, and a source of much difference of opinion and expensive litigation would at once be removed. Of the cost of enclosing and maintaining such reclaimed lands, and their success as speculations, I am not enabled from any experience of my own to judge. But, referring to what has already been said at page 321, I can safely say that unless the surface of the marsh to be enclosed is on a high level, it is not expedient to enter on works for its reclamation.

Even after enclosure the embankments have to be attended to, kept in repair at a constantly recurring expenditure, and often additional works have to be employed for further protection; and I have still shortly to notice some of the protection structures that have been erected in defending the banks of rivers and shores of estuaries.

In Holland, as is well known, the reclamation and protection of land, both from the sea and from rivers, has been carried to a greater extent than in any other country, and much useful information will be found on that subject, and indeed on reclamation generally, in the papers by Mr. Paton, Mr. Oldham, and Mr. J. H. Muller, in the *Proceedings of the Institution of Civil Engineers.*[1]

There can be no doubt that a smooth surface tends

[1] Vol. xxi.

to preserve the banks of a river. The water having no obstruction glides gently past without disturbance. But if the river's banks have, from neglect, got into a rugged, uneven state, I have found that a very sluggish stream may produce an abrading action in excess of what its velocity seemed to warrant. The rugged outline of the bank produces on a small scale the effect described at page 176 as resulting from jetties. The projecting points of grass-covered alluvial soil act as so many obstructions to the

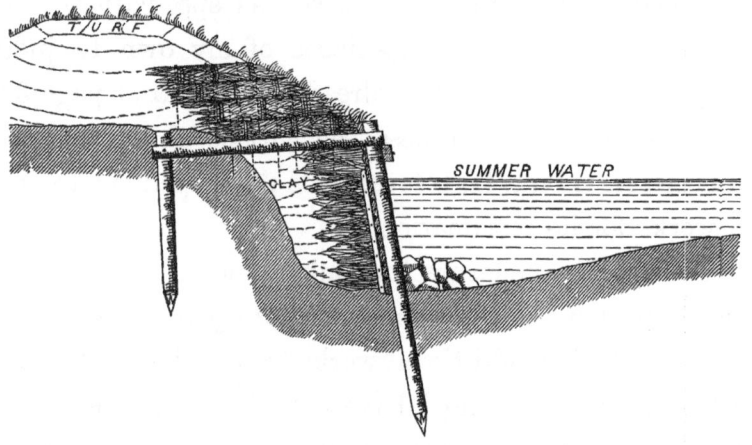

Fig. 64.

current, and in such a case the abrading action of the river cannot be measured by the general velocity of the stream, but by the *local* velocity (if I may use the expression) with which it sweeps round, and gradually undermines the rugged parts of the bank. Although the passage of a float down the centre of a stream indicated a velocity too slow to abrade a river's bank, it would be erroneous to assume that therefore there are no local currents round the salient points of the foreshore strong enough to wear them away.

Sometimes stones are deposited to cover gently sloping banks, and where they are steep I have found piling and brushwood, arranged as shown in fig. 64, a very effectual protection for rivers having winding courses and soft beds.

In other cases, in more open estuaries exposed to the Works for protection of land
sea, works of a stronger kind are required.   Figs. 65 and in open
estuaries.
66 are a plan and section of a protection which was used

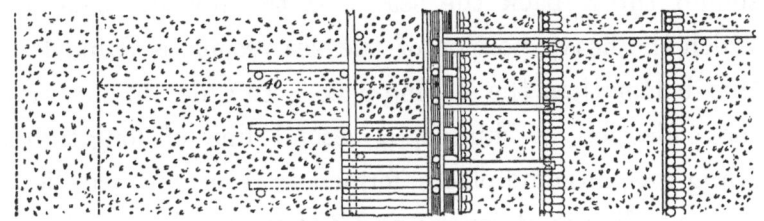

FIG. 65.

on a line of shore composed of shingle.   Jetties projecting from the shore had at first been used to collect the shingle,

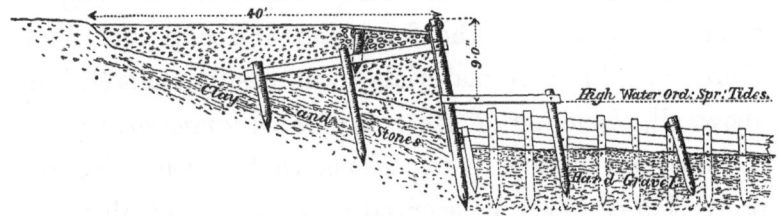

FIG. 66.

but I found that in heavy seas the waves were led along the jetties, and had a hurtful effect at their roots where they joined the beach.   A continuous line of piling and planking was accordingly adopted, combined with occasional jetties, as shown above, and this has proved very successful.   In proof of this, it has been found that wherever the upright piling and planking has been formed, there was no influx of anything beyond spray upon the adjoining land, but that at all other parts of the coast (which is about 6 miles in length), where the face of the beach is sloping, the water passed freely over in considerable depth, carrying drift timber far into the fields, and in some places heavy shingle to the depth of 2 feet. The problem to be solved was to oppose an obstacle which should throw back the sea; and the upright face, from which the heavy portion of the sea recoils, is found to do this better than the sloping face. In order to encourage the collection of shingle, a second line of longitudinal piling was, at some places, formed in front, and parallel to the main line of defence ; and the works described have been found a very effective defence on a line of shingle beach, exposed to a considerable sea, on the shores of the Bristol Channel.

In designing all such works, however, the engineer must be guided by the formation and exposure of the shores, the kind of materials most easily available, and, above all, *the value of the property endangered*, as every engineer must know by experience that in some situations protection can only be secured at a cost out of all proportion to the benefit which it would confer.

CHAPTER XIV.

CROSSING OF NAVIGATIONS BY RAILWAY BRIDGES.

THE interests of railway companies and the conservators of navigations are often antagonistic. It is not unfrequently an object of great importance for a railway company to obtain a crossing over a navigable river, and they not unnaturally, in their zeal to promote the interests of their shareholders, undervalue the importance of navigation. The consequence is, that many of the hardest Parliamentary battles have of late years originated in the conservators of navigable rivers resisting the attempts of railway companies to carry out their schemes with little regard to the obstruction they may interpose to sea-borne traffic, or the ruin they may entail on old-established trade.

The question as to the propriety of permitting a railway to cross a navigation, must obviously depend on the relative importance of the railway and river traffic—the amount of interference proposed—the interests of those connected with the two trades, and many other points which it is neither my province nor intention to discuss in this place. But so much has the question of *bridging* navigable rivers of late been brought under notice, that it seems desirable to lay before the reader a statement of

the *general* grounds on which such interferences have
been opposed on behalf of the interests of navigation, as
being a not unsuitable topic in connexion with the subject
of river improvements which we have been considering.

It is not my intention to refer to schemes for crossing
rivers by high-level bridges of great span, but to the more
general interference caused by railways crossing on a low
level, with opening spans for the passage of vessels.

It is both natural and necessary that the conservators
of the public highways, afforded by rivers, should look
with no friendly eye on any attempt to obstruct or
injure their usefulness. The public owe much to the
firm, and in many cases successful, opposition offered
on their behalf to some schemes of railway companies
designed to cross a navigable river, in order to save a
few miles' detour or avoid using part of the line belonging
to another company. I do not by any means say that
all railway crossings are, or have been, of this character.
There are, and may again be, cases where the benefit
derived by the public from a railway bridge across a
river, so greatly outweighs any benefit that can possibly
be derived from preventing its erection, that the navigation may freely yield to the railway. But this is not
always the case; sometimes the two interests may be
fairly balanced, and in other cases the proposal to bridge
a navigable river cannot for a moment be entertained.
No one at present would dream of interposing a railway
crossing and swing-bridge between London and Green-
wich, or between Glasgow and Greenock.

The objections to such crossings have generally been

argued from two distinct points of view, the one founded on *nautical*, and the other on *engineering* grounds.

The nautical question refers to the mode of navigating our tidal rivers, and the difficulty of taking vessels through the narrow opening of a swing-bridge in a rapid tide-way. The arguments adduced are, that our rivers are entirely dependent on the flow of the tides, and are navigable for ships, *only* when the tidal water is in the channel. At low water they are shallow fresh-water streams, sometimes navigable only by small boats. The time for the passage of large vessels is restricted to one or two hours before and after high water, and it is absolutely necessary for vessels to take advantage, not only of high water, but of the best tides, both in making and leaving the ports on rivers. Every obstruction, therefore, that may tend to hinder the progress of a vessel, and lead to her losing a tide, is a very serious evil, and renders it desirable that no obstacle should be placed in her course. Unless vessels can run freely in and out, they will not continue to frequent a port.

For the same reasons objections have been raised to the control which such a crossing places in the hands of the company. The opening of the bridge must be so regulated as to suit the passage of trains, which is, or ought to be, *regular*, whereas the time of high water varies from day to day. These, and other objections, have been often advanced to show the incompatibility of the two interests.

The engineering objections are founded mainly on the fact that piers placed in the water-way of a river disturb the currents and cause shoals, as will be best understood by referring to page 190, where the effects of such disturbances are fully discussed. If the bed of the river is composed of rock or other hard material, the objection founded on shoaling ceases ; but in a river having a bed of gravel, or, still worse, of sand, the case is very different. The amount of scour will vary with the state of the tides and the amount of *fresh* in the river, and shoals must necessarily be thrown up, varying in position and amount according to the currents which produced them. These obstructions, as we have seen, may be caused by a single tide, so that no application of dredging can remove the objection in sufficient time to restore the navigable depth of channel for passing vessels. It is well known that the *shoalest water* becomes the *ruling* depth of the navigation, and a shoal which reduces the depth by one foot, only at one point, practically reduces the depth for navigation by that amount. So that if, after high spring-tides or a heavy flood in the river, a shoal is caused above or below the bridge, it becomes a formidable impediment, all the more so that the railway company, with the very best intention, can do nothing to remedy the evil, which may spring up in a single night. This introduction of any element of un-certainty as to depth is perhaps the greatest evil that can be inflicted on the interests of a port.

An artificial covering for the bed and banks of the river, similar to what has been described in Chapter VIII., may perhaps be suggested as a remedy. But such a covering would require to extend for a great way on

PLATE XVI.

RIVER OUSE SWING BRIDGE.

From the Proceedings of the Institution of Mechanical Engineers, Birmingham.

GENERAL ELEVATION.

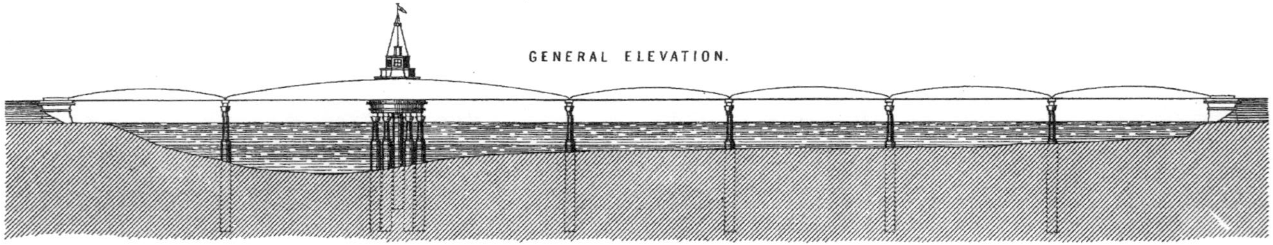

P L A N.

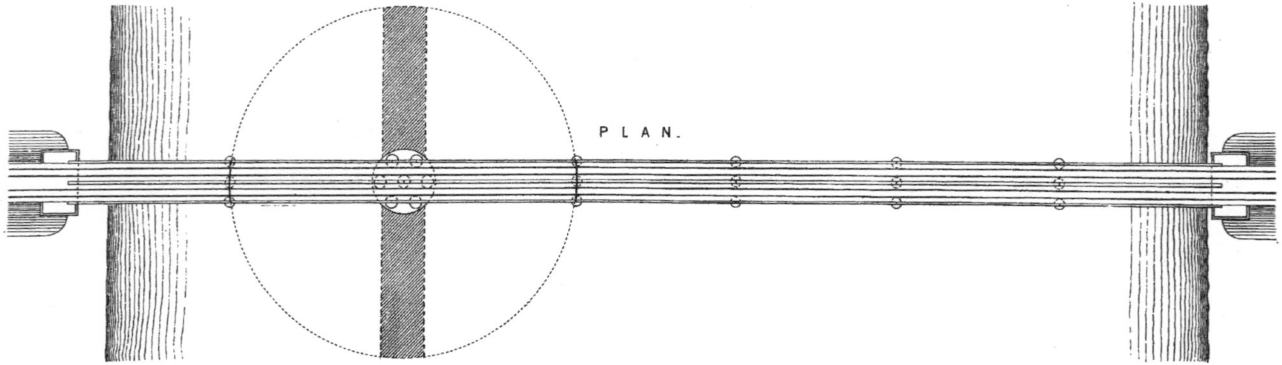

SCALE

100   50   0   100   200   300 Feet.

J.Bartholomew, Edin.r

Published by A. & C. Black, Edinburgh.

either side of the bridge, and would be enormously costly, whatever the material employed. It would prevent further deepening of the river, and no vessel could drop anchor on it. Above all, it might be found that the sudden change from a hard to a soft bed at either extremity produced disturbance of currents and shoals as inconvenient as those caused by the piers of the bridge which it was designed to prevent.

Swing-bridges have been sanctioned on several navigable rivers, and attempts to erect them on others have been successfully resisted. The railway authorities have invariably admitted that if such crossings are allowed, they should be arranged so as to be as little injurious as possible ; and as I believe the hydraulic swing-bridge, designed for crossing the Ouse near Goole by Mr. T. E. Harrison, is the most perfect structure of the kind that has been made, I shall give a short description of its leading features, referring the reader for details to the elaborate description and drawings communicated to the Institution of Mechanical Engineers, by Sir William Armstrong.

It will be seen from Plate XVI., which is from Sir W. Armstrong's paper, that the bridge has seven spans. The main pier, which is about 40 feet in width and 250 feet in length, is placed in the deepest part of the river. On this pier the moveable part of the bridge revolves, forming, when it is open, a passage for vessels of about 100 feet in width on each side of the pier.

The following are some of the general descriptions and dimensions, from the paper referred to :—

" The total length of the bridge, fixed and moveable,

is 830 feet. The fixed portions consist of five spans, of 116 feet each from centre to centre of piers. Each of the piers for the fixed spans consists of three cast-iron cylinders of 7 feet diameter, and about 90 feet length. The depth from the under side of the bridge to the bed of the channel in the deepest part is about 61 feet. The headway beneath the bridge is 14 feet 6 inches from high-water datum, and 30 feet 6 inches from low-water. The swing portion of the bridge consists of three main wrought-iron girders, 250 feet long, and 16 feet 6 inches deep at the centre, diminishing to 4 feet deep at the ends. The centre girder is of larger sectional area than the side girders, and instead of being a single web, is a box girder 2 feet 6 inches in width. An annular box girder, 32 feet mean diameter, is situated below the centre of the bridge, and forms the cap of the centre pier. This girder is 3 feet 2 inches in depth, and 3 feet in width, and rests upon the top of six cast-iron columns, each 7 feet diameter, which are arranged in a circle, and form the centre pier of the bridge. Each of these columns has a total length of 90 feet, being sunk about 29 feet deep in the bed of the river. A centre column, 7 feet diameter, is securely braced to the six other columns by a set of cast-iron stays, which support the floor of the engine-room. This centre column contains the accumulator, and forms the centre pivot for the rotation of the bridge.

"The weight of the swing bridge is 670 tons, and rests upon a circle of conical rollers. These are twenty-six in number, each 3 feet diameter, with 14 inches width of tread, and made of cast-iron hooped with steel. They run between the two circular roller-paths, 32 feet diameter, and 15 inches broad, which are made of cast iron, faced with steel; the axles of the rollers are horizontal, and the two roller-paths are turned to the same bevel.

" The turning motion is communicated to the bridge by means of a circular cast-iron rack, $12\frac{1}{2}$ inches wide on the face and $6\frac{1}{2}$ inches pitch, which is shrouded to the pitch line, and is bolted to the outer circumference of the upper roller-path. The rack gears with a vertical bevel wheel, which is driven by a pinion connected by intermediate gearing with the hydraulic engine. There are two of these engines, duplicates of one another, and either of them is sufficient for turning the bridge, the force required for this purpose being equal to about ten tons applied at the radius of the rollers' path. Each hydraulic engine is a three-cylinder oscillating engine, with simple rams of $4\frac{1}{4}$ inches diameter and 18 inches stroke. These engines work at forty revolutions per minute, with a pressure of water 700 lbs. per inch, and are estimated at forty horse-power each. The steam-engines for supplying the water-pressure are also in duplicate, and are double-cylinder engines, driving three throw pumps of $2\frac{3}{4}$ inches diameter and 5 inches stroke, which deliver into the accumulator. The steam cylinders are 8 inches diameter and 10 inches stroke, each engine being twelve horse-power. The accumulator has a ram $16\frac{1}{2}$ inches diameter, with a stroke of 17 feet. It is loaded with a weight of 67 tons."

There are other interesting details, showing the highly

ingenious mechanism designed by Mr. Harrison and Sir W. Armstrong, for adjusting the bridge so as to obtain a perfectly solid roadway, and for other arrangements connected with the railway traffic, into which I need not enter. Sir William closes his paper by stating that "the time required for opening or closing the bridge, including the locking of the ends, is only fifty seconds, the average speed of motion of the bridge-ends being 4 feet per second. For the purpose of insuring safety in the working of the railway line over the bridge, a system of self-acting signals is arranged, moved by the fixing gear at the two ends of the bridge; and a signal of 'all right' is shown by a single semaphore and lamp at each end of the fixed parts of the bridge; but this cannot be shown until each one of the locking bolts and resting blocks is secure in its proper place."

It has sometimes been proposed to cross navigations by carrying the railway *below* the bed of the river; and if this could be done by *tunnelling*, the objections to which allusion has been made would be entirely obviated. But the schemes brought before Parliament for *under-water* crossings have invariably been coupled by a proposal to execute the works by *open cutting*, either by diverting the river or constructing cofferdams in its bed. The difficulty of dealing in this manner with a large river, without, for a considerable period, seriously obstructing its navigation, and the obvious disadvantages of such a crossing, both as regards gradients and drainage, have hitherto, so far as I know, prevented any such scheme being passed by the Legislature.

PHYSICAL CHARACTERISTICS OF RIVERS.

Name of River.	Length in miles.	Area of drainage in square miles.	Ordinary discharge per minute in cubic feet.	No. of cubic feet discharged per square mile of drainage per minute.	Slope of surface in inches per mile.	Part of river where slope occurs.	Length of river affected by tide in miles.	Depth on bar at low water in feet.	Authority.
Amazon, ...	4000	...	...	...	2·34	......	400	...	N.Breadmore's Hyd.Tabls.
Annan, ...	35	...	...	...	66	{ Annan Waterfoot to Annan Bridge, 2 miles, ... }	2	...	{ Messrs. Stevenson, C.E., Edinburgh.
Boyne, ...	60	700	180,000	257	...	......	2	...	A. Nimmo, C.E.
Clyde, ...	98	945	48,000	65	1⅓	{ Broomielaw to Port-Glasgow, 18 miles, ... }	22 above Port-Glasgow.	} no bar	J. Ure, C.E, Glasgow.
Conon, ...	35	399	7,969	19·9	...	......		...	Messrs. Stevenson, Edin.
Coquet, ...	44	...	...	...	60	......	3¼	...	E. K. Calver, R.N.
Danube to Isakteha, ...	1700	300,000	7,500,000	25	5	Ismail Chatal to sea, ...		15to17½	Sir C. A. Hartley, C.E.
Dee, Aberdeen, ...	87	765	10,675	...	81	Lower part, ...	...	...	J. Gibb.
Dee, Chester, ...	85	620	...	...	11	Chester to Flint, ...	32	10 to 12	{ Messrs.Stevenson,Edin., and Admiralty report, by Capt. Washington.
Forth, ... Before works were executed,	63 Above Alloa.	452	...	75·7	11 13	{ Black Dub to Stirling ... { Stirling to Alloa, ...	15 above Alloa.	...	Messrs. Stevenson.
Foyle, ...	55	1100	31,500	28·6	1·25	{ Londonderry to Culmore Point, ...	23 above Londonderry.	...	Messrs. Stevenson.
Ganges, ...	1680	432,480	12,420,000	28·7	3·37	{ Rajmahal to Mirzapore Creek, ...	...	...	{ Johnston's Physical Atl. Beardmore's Tables. Rev. Mr. Everest.
Irrawady, ...	...	...	4,500,000	...	1·6 3·8	{ in summer, ... { in flood, ...	105	...	T. Login, C.E.
Lune, ...	50	...	...	...	26·76	{ Glasson to Heaton, 3¼ miles, ...	...	...	Messrs. Stevenson.

Y

PHYSICAL CHARACTERISTICS OF RIVERS—*Continued.*

Name of River.	Length in miles.	Area of drainage in square miles.	Ordinary discharge per minute in cubic feet.	No. of cubic feet discharged per square mile of drainage per minute.	Slope of surface in inches per mile.	Part of river where slope occurs.	Length of river affected by tide, in miles.	Depth on bar at low water in feet.	Authority.
Lune—Before works were executed,	...	...	...	...	20· / 24·	Heaton to Lancaster, 2¼ / Average slope between Glasson and Lancaster,	...	...	Messrs. Stevenson.
Mersey,	70	1,748	...	...	......		...	9 to 10	Captain Denham, R.N.
Mississippi,	4400 including Missouri.	1,226,600	39,954,000	32·5	2·8 / 3·25	ordinary {from Ohio to Gulf / flood, ...{ of Mexico......	...	...	C. Ellet.
Ness,	7	700	...	...	96	Loch Dochfour to Kessock,	...	...	Messrs. Stevenson.
Nile,	2240	520,200	1,386,000	2·6	5·5 / 3·25	when high { Cairo to Medi- / when low,{ terranean,	2	...	N. Beardmore. / Johnston's Physical Atl.
Nith,	45	...	...	...	20	Dumfries to Burron Point,	...	...	Messrs. Stevenson.
Rhine,	700	88,853	3,960,000	44·5	2119 / 119 / 63 / 20·7 / 7·7	Source to Reichenau, / Reichenau to Constance, / Constance to Basle, / Basle to Cologne, / Cologne to sea,	...	...	Johnston's Physical Atl. / Beardmore's Tables. / Falls and Discharge, from L. Horner, Esq.
Rhone,	560	38,329	...	...	24·18	Besancon to Mediterranean,	...	...	Johnston's Physical Atl. / Beardmore's Tables.
Ribble, before works,	80	880	139,935	159	47·8	Lower 22½ miles,	22½	5½	P. Park, C.E., Preston.
Severn,	180	8580	...	...	4·5 / 2 / 3 / 2 / 0·75	Diglis Weir to Upton Br., / Upton Br. to Mythe Br., / Myth Bridge to Haw Br., / Haw Br. to Stonebench, / Stonebench. to Rosemary Pt.,	68	...	Remarks on the Tidal Phenomena of the river Severn, by Capt. F. W. Beechey, R.N.

PHYSICAL CHARACTERISTICS OF RIVERS—*Continued.*

Name of River.	Length in miles.	Area of drainage in square miles.	Ordinary discharge per minute in cubic feet.	No. of cubic feet discharged per square mile of drainage per minute.	Slope of surface in inches per mile.	Part of river where slope occurs.	Length of river affected by tide, in miles.	Depth on bar at low water in feet.	Authority.
Severn, ...	180	8580	...	...	5 10·5 7·5 15 20·5 23·25 7	Rosemary Point to Framilode, ... Framilode to Newnham, Newnham to Awre Point, Awre Point to Sharpness, Sharpness to Guscar, ... Guscar to Inward Point, Inward Pt. to Aust Head,	68	...	Remarks on the tidal Phenomena of the river Severn, by Capt. F. W. Beechey, R.N.
Tay, before works, ...	160	2283	274,000	120	1·95 9·35 5·06	Flisk to Balmbreich, ... Balmbreich to Newburgh, Newburgh to Perth, ...	35	16 to 18	Messrs. Stevenson.
Tees, ...	100	700	...	...	...	....	27	ft. in. 3 to 12·6 aver. 9f.	John Fowler, C.E.
Thames, ...	204½ Navigable.	5000	102,000	20·4	20 ·92	Lechdale to Teddington, Teddington to Yanlet Crk.,	66 from Nore.	no bar	Messrs. Rennie, C.E.
Tweed, ...	100	1870	...	...	109 116 49	Confluence of Gala Water to Leader Water, ... Leader to Kelso, ... Kelso to sea, ...	...	...	A. Peterman, F.G.S.
Tyne, ...	80	1142	...	...	180 84 45 $\frac{6}{10}$ 10 $\frac{1}{3}$	Source to Hexham, ... Hexham to Stocksfield, Stocksfield to Ryton, ... Ryton weir to Bar, ...	19½	6 to 7	W. A. Brooks, C.E.
Wear, ...	58	437	9,500	22	16	New Bridge to sea, ...	10	3 to 4	Th. Meik, C.E., Sunderland.

INDEX.

PRINTED BY T. AND A. CONSTABLE, PRINTERS TO HER MAJESTY,
AT THE EDINBURGH UNIVERSITY PRESS.

In One Volume 8vo, Illustrated with Woodcuts and Plates, Price 10s. 6d.

HARBOURS

THEIR DESIGN AND CONSTRUCTION.

BY

THOMAS STEVENSON, F.R.S.E., M.I.C.E.

"This work is undoubtedly of great value."—*The Builder.*

"A most valuable addition to the Engineers' library. The chapters devoted to 'Generation of Waves,' and to 'Force of Waves,' are admirable treatises."—*The Artizan.*

"We know of no work from which an engineering student is likely to derive more solid advantage than from this. Had we space we should gladly refer to some of the more striking parts of the work; we can, however, only refer to it as a whole, with unmixed commendation."—*Practical Mechanics' Journal.*

"Mr. Stevenson has given us so much valuable matter in a condensed form in this book that we have not attempted to give anything like a synopsis of its contents; to the student it will prove especially valuable."—*Mechanics' Magazine.*

"We commend it to the careful and attentive perusal of all connected with the construction of harbours, docks, and other kindred works of marine engineering."—*Newcastle Journal.*

"The present work, by one who possesses a most thorough and scientific knowledge of his subject, is likely to prove a valuable contribution to marine engineering."—*Glasgow Herald.*

"This contribution to the library of the civil engineer may be considered as a series of important facts, which should guide him in all his proceedings."—*Nautical Magazine.*

EDINBURGH: ADAM AND CHARLES BLACK.

Just Published, in Demy 8vo, Second Edition, Price 12s. 6d.

LIGHTHOUSE ILLUMINATION

BEING A DESCRIPTION OF THE HOLOPHOTAL SYSTEM, AND OF AZIMUTHAL, CONDENSING, AND OTHER NEW FORMS OF LIGHTHOUSE APPARATUS.

BY

THOMAS STEVENSON, F.R.S.E., M.I.C.E.,

Illustrated with numerous Woodcuts and Plates.

Notices of First Edition.

"It is no mean merit to have advanced beyond so illustrious a predecessor as Fresnel, but we conclude the perusal of the clear and interesting account before us, with the conviction that modern engineers of lighthouses, and Mr. Stevenson pre-eminently among them, have succeeded in several important particulars in improving and extending Fresnel's method. . . . The present work, besides being a summary account of established modern improvements of lighthouses, is valuable as a lucid and clear exposition of the more recent modifications of the author's valuable system of illumination."—*Civil Engineers' and Architects' Journal.*

"In the present instance, Mr. Thomas Stevenson has well reminded us that he has not forgotten his combined engineering and literary craft; for, whilst we can all of us go and examine his finished works, we can also, as the volume before us now tells us, turn to a scientifically arranged book, wherein the whole system of 'lighthouse illumination' is fully and carefully explained. . . . The book is one which tells all that is at present known on the difficult subject of applied optics for the transmission of light to great distances."—*The Practical Mechanics' Journal.*

"The perfect manner in which every particle of light is used in these days is in good keeping with the improvement which is everywhere seen. The object of the present Treatise is to explain the principles of the new system; by which, as is shown by the author, the utmost can be obtained from a single source of light. We commend this highly approved invention of Mr. Thomas Stevenson, with the valuable treatise in Weale's Library of Mr. Alan Stevenson, to the attention of those who wish for the best system of lighthouse illumination."—*The Nautical Magazine.*

EDINBURGH: ADAM AND CHARLES BLACK.